# 411 SAT

# ALGEBRA AND GEOMETRY QUESTIONS

# 411 SAT

## ALGEBRA AND GEOMETRY QUESTIONS

LEARNINGEXPRESS ®

NEW YORK

Library of Congress Cataloging-in-Publication Data:
411 SAT algebra and geometry questions.
    p. cm.
  ISBN 1-57685-560-0
1. Algebra—Examinations, questions, etc.  2. Geometry—Examinations, questions, etc.
3. SAT (Educational test)—Study guides.  I. LearningExpress (Organization).
II. Title: Four hundred eleven SAT algebra and geometry questions.
QA157.A14 2006
512.0076—dc22                                                    2006005374

Printed in the United States of America

9 8 7 6 5 4 3 2 1

ISBN 1-57685-560-0

For information on LearningExpress, other LearningExpress products, or bulk sales, please write to us at:
  LearningExpress
  55 Broadway
  8th Floor
  New York, NY 10006

Or visit us at:
  www.learnatest.com

# Contents

# CONTENTS

# 411 SAT
## ALGEBRA AND GEOMETRY QUESTIONS

# Introduction ▶

I hate math." "Math is my worst subject." "I'll never understand math—it's too hard."

So many people approach math with an attitude that dooms them to failure before they even get started. Your approach to math, or any subject, can make all the difference in the world. With the majority of SAT math questions involving algebra or geometry, this book is designed not only to give you the algebra and geometry skills you need for the exam, but also to provide you with the confidence you require to score highly on the exam.

You can't study every possible word that may appear on the verbal portion of the SAT. No such list even exists. However, all of the math skills that you will encounter on the math portion of the SAT are known, and the most common of them are in this book. In this respect, it's actually easier to study for the math sections of the SAT than the verbal sections.

This book trains you for the SAT with 411 math questions. Every question can be solved using the skills described in the following 16 chapters, and every answer is explained. If an explanation doesn't make sense, review the related material in the chapter, and then try the question again.

Don't get discouraged. Use these chapters to gain an understanding of the algebra and geometry that's given you trouble in the past. With time and practice, you'll become comfortable with math and enter the SAT with confidence, believing you are prepared for any question.

## ▶ An Overview of This Book

This book is divided into an algebra section and a geometry section, each comprised of eight chapters, plus a pretest and a posttest. The tests are made up of only algebra and geometry questions. Use the pretest to identify the topics in which you need improvement. Use the posttest to help you identify which topics you may need to review again. The first eight chapters focus on algebra skills:

### Chapter 1: Algebraic Expressions
Understand the parts of algebraic expressions, combine like and unlike terms, evaluate expressions using substitution, and evaluate one variable in terms of another.

### Chapter 2: Solving Equations and Inequalities
Use basic arithmetic and cross multiplication to solve single- and multi-step equations, find the values that make an expression undefined, and form algebraic equations from word problems.

### Chapter 3: Quadratic Expressions and Equations
Multiply binomials, factor and solve quadratic equations, and analyze the graphs of quadratic equations.

### Chapter 4: Factoring and Multiplying Polynomials
Multiply, factor, and find the roots of polynomials.

### Chapter 5: Radicals and Exponents
Add, subtract, multiply, divide, and simplify radicals and terms with exponents; rationalize denominators and solve equations with radicals. Work with negative and fractional exponents, and raise exponents to exponents.

### Chapter 6: Sequences
Solve arithmetic, geometric, and combination sequences for the next or missing term.

### Chapter 7: Systems of Equations
Solve systems of two equations with two variables using substitution and combination.

### Chapter 8: Functions, Domain, and Range
Determine if an equation is a function using the vertical line test, find the domain and range of functions, solve nested functions and evaluate functions with newly defined symbols.

The last eight chapters focus on geometry skills:

### Chapter 9: Angles
Recognize and use the properties of acute, obtuse, right, straight, complementary, supplementary, vertical, and alternating angles.

### Chapter 10: Triangles

Recognize and use the properties of interior and exterior angles of triangles and acute, obtuse, right, scalene, isosceles, equilateral, and similar triangles.

### Chapter 11: Right Triangles

Use the Pythagorean theorem to find the missing side of a triangle. Review the properties of special right triangles: 45-45-90 (isosceles) right triangles and 30-60-90 right triangles, and use basic trigonometry to determine the size of angles and sides of right triangles.

### Chapter 12: Polygons

Review the properties of interior and exterior angles of polygons and regular polygons. Find the perimeter of polygons and work with similar polygons.

### Chapter 13: Quadrilaterals

Learn the differences that distinguish parallelograms, rhombuses, rectangles, and squares from each other, and the similarities that each share.

### Chapter 14: Area and Volume

Find the area of triangles and rectangles (including squares), the volume of cylinders and rectangular solids (including cubes), and the surface area of solids.

### Chapter 15: Circles

Review the parts of a circle, including radius and diameter, and use them to find the circumference and area of a circle, as well as the area of a sector of a circle and the length of an arc of a circle.

### Chapter 16: Coordinate Geometry

Plot and find points on the coordinate plane; find the slope, midpoint, and distance of line segments.

## ▶ How to Use This Book

Start at the beginning. Each chapter builds on the skills reviewed in the chapters that precede it. There's a reason the algebra chapters come first—the SAT is filled with algebra questions, and many of the geometry questions you'll encounter will involve algebra. The last eight chapters focus on geometry, but you'll need the algebra skills reviewed in the first eight chapters to solve these problems.

In each chapter, a set of skills is reviewed (with key words defined), including examples of each skill with explanations. Following the lesson portion of the chapter are 15–30 SAT-caliber questions. Related questions are grouped together. In general, the questions are progressively more difficult. If you can handle the questions in each chapter, you're ready for the SAT!

Read and reread each section of this book. If you don't understand an explanation of a geometry question that involves algebra, it may be helpful to reread an algebra section—you might have the geometry skill mastered, but you may be confused about the algebra part of the question.

Algebra and geometry questions comprise the bulk of SAT math, but this book does not cover every type of math question you'll see on the SAT. Review a book such as *SAT Math Essentials* or *Acing the SAT 2006* by LearningExpress to be sure you've got all the skills you need to achieve the best possible math score on the SAT. Good luck!

# Pretest ▶

**B**efore you begin Chapter 1, you may want to get an idea of what you know and what you need to learn. The pretest will answer some of these questions for you. The pretest consists of 25 questions that cover the topics in this book, of which 15 are the multiple-choice questions and 10 are the grid-ins. For the grid-ins, you come up with the answer yourself instead of choosing from a list of possible answers. While 25 questions can't cover every concept, skill, or shortcut taught in this book, your performance on the pretest will give you a good indication of your strengths and weaknesses. Keep in mind the pretest does not test all the skills taught in this book, but it will tell you the degree of effort you will need to put forth to accomplish your goal of learning algebra and geometry.

If you score high on the pretest, you have a good foundation and should be able to work your way through the book quickly. If you score low on the pretest, don't despair. This book will take you through the algebra concepts, step by step. If you get a low score you may need to take more than 20 minutes a day to work through a lesson. However, this is a self-paced program, so you can spend as much time on a lesson as you need. You decide when you fully comprehend the lesson and are ready to go on to the next one.

Take as much time as you need to do the pretest. When you are finished, check your answers with the answer key at the end of the book.

## PRETEST

1. ⓐ ⓑ ⓒ ⓓ ⓔ
2. ⓐ ⓑ ⓒ ⓓ ⓔ
3. ⓐ ⓑ ⓒ ⓓ ⓔ
4. ⓐ ⓑ ⓒ ⓓ ⓔ
5. ⓐ ⓑ ⓒ ⓓ ⓔ

6. ⓐ ⓑ ⓒ ⓓ ⓔ
7. ⓐ ⓑ ⓒ ⓓ ⓔ
8. ⓐ ⓑ ⓒ ⓓ ⓔ
9. ⓐ ⓑ ⓒ ⓓ ⓔ
10. ⓐ ⓑ ⓒ ⓓ ⓔ

11. ⓐ ⓑ ⓒ ⓓ ⓔ
12. ⓐ ⓑ ⓒ ⓓ ⓔ
13. ⓐ ⓑ ⓒ ⓓ ⓔ
14. ⓐ ⓑ ⓒ ⓓ ⓔ
15. ⓐ ⓑ ⓒ ⓓ ⓔ

16.    17.    18.    19.    20.

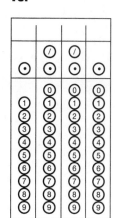

21.    22.    23.    24.    25.

**1.** The expression $\frac{(9x^2 + 9y^2)}{(6x^2 + 20x + 6)}$ is undefined when

    **a.** $x = 0$ or $y = 0$

    **b.** $x = \frac{1}{3}$ or $y = -\frac{1}{3}$

    **c.** $x = 1$ or $y = 6$

    **d.** $x = -\frac{1}{3}$ or $y = -3$

    **e.** $x = -1$ or $y = -6$

**2.** Each term in the sequence below is five times the previous term. What is the eighth term in the sequence?

    4, 20, 100, 500, ...

    **a.** $500 \times 8$

    **b.** $4^8$

    **c.** $4 \times 5^7$

    **d.** $4 \times 5^8$

    **e.** $5 \times 4^8$

**3.** A circle has an area of $64\pi$ ft.$^2$. What is the circumference of the circle?

    **a.** 8 ft.

    **b.** $8\pi$ ft.

    **c.** $16\pi$ ft.

    **d.** $32\pi$ ft.

    **e.** 32 ft.

**4.** Which of the following is true of $y = 2$?

    **a.** It is not a function.

    **b.** It has a range of 2.

    **c.** It has no domain.

    **d.** It has a slope of 2.

    **e.** It has no $y$-intercept.

**5.** What is the value of $(\frac{2a}{b})(\frac{a^{-1}}{(2b)^{-1}})$?

    **a.** 1

    **b.** 2

    **c.** $2a$

    **d.** 4

    **e.** $\frac{a^2}{b^2}$

**6.** The inequality $-4(x - 1) \leq 2(x + 1)$ is equivalent to

    **a.** $x \geq -\frac{1}{3}$

    **b.** $x \leq -\frac{1}{3}$

    **c.** $x \geq \frac{1}{3}$

    **d.** $x \leq \frac{1}{3}$

    **e.** $x \leq 3$

**7.** Side $AB$ of rectangle $ABCD$ measures 10 units. Side $BC$ of the rectangle is shared with side $BC$ of square $BCEF$. If the area of square $BCEF$ is 20 square units, what is the area of rectangle $ABCD$?

    **a.** $10\sqrt{5}$ square units

    **b.** $20\sqrt{5}$ square units

    **c.** $20 + 20\sqrt{5}$ square units

    **d.** 100 square units

    **e.** 200 square units

**8.** A circular compact disc has a diameter of 12 cm and a height of 2 mm. If Kerry stacks 10 compact discs on top of each other, what is the volume of the stack?

    **a.** $72\pi$ cm$^3$

    **b.** $144\pi$ cm$^3$

    **c.** $240\pi$ cm$^3$

    **d.** $288\pi$ cm$^3$

    **e.** $720\pi$ cm$^3$

**9.** Which of the following statements is true?

    **a.** All squares are rectangles and rhombuses.

    **b.** All rectangles are rhombuses, but not all rhombuses are rectangles.

    **c.** All rhombuses are parallelograms and all parallelograms are rhombuses.

    **d.** All rhombuses are squares, but not all squares are rhombuses.

    **e.** All squares are parallelograms, but not all squares are rectangles.

**10.** For what values of $x$ is the expression $3x^2 - 3x - 18$ equal to 0?

   **a.** $x = -3, x = 2$

   **b.** $x = -3, x = 6$

   **c.** $x = 3, x = -\frac{1}{2}$

   **d.** $x = 3, x = -6$

   **e.** $x = 3, x = -2$

**11.** If $ABCD$ (shown below) is a parallelogram, what is the measure of angle $ABC$?

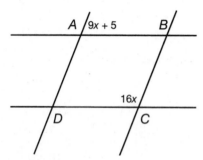

   **a.** 7°

   **b.** 68°

   **c.** 78°

   **d.** 102°

   **e.** 112°

**12.** Triangles $ABC$ and $DEF$ are similar. Side $AB$ of triangle $ABC$ corresponds to side $DE$ of triangle $DEF$. If the length of $\overline{AB}$ is 10 units and the length of $\overline{DE}$ is 4 units, then the area of triangle $DEF$ is equal to

   **a.** $\frac{4}{25}$ (area of triangle $ABC$).

   **b.** $\frac{1}{5}$ (area of triangle $ABC$).

   **c.** $\frac{2}{5}$ (area of triangle $ABC$).

   **d.** $\frac{5}{2}$ (area of triangle $ABC$).

   **e.** $\frac{25}{4}$ (area of triangle $ABC$).

**13.** Which of the following parabolas has its turning point in the second quadrant of the coordinate plane?

   **a.** $y = (x + 1)^2 - 2$

   **b.** $y = (x - 1)^2 - 2$

   **c.** $y = -(x + 1)^2 - 2$

   **d.** $y = -(x + 2)^2 + 1$

   **e.** $y = (x - 2)^2 + 1$

**14.** If $a = \frac{7b - 4}{4}$, then $b =$

   **a.** $\frac{a}{7}$

   **b.** $\frac{4a}{7}$

   **c.** $\frac{a + 1}{7}$

   **d.** $\frac{4a + 4}{7}$

   **e.** $\frac{7a - 4}{7}$

**15.** Jennifer makes a hexagon by arranging 6 equilateral triangles, each of which has a vertex at the center of the hexagon. If the length of one side of a triangle is 4, what is the area in square units of the hexagon?

   **a.** $6\sqrt{3}$

   **b.** 12

   **c.** $12\sqrt{3}$

   **d.** 24

   **e.** $24\sqrt{3}$

**16.** If the surface area of a cube is 96 square centimeters, what is the volume of the cube in cubic centimeters?

**17.** What is the distance between the points $(-7, -4)$ and $(5, 12)$?

**18.** Steve draws a polygon with 12 sides. What is the sum of the measures of the interior angles of Steve's polygon?

**19.** Find the value of $\frac{\sqrt{b+13}}{b} \, b\sqrt{b}$ if $b = 36$.

**20.** In the diagram below, sides $AC$ and $AB$ of triangle $ABC$ are congruent. If the measure of angle $DCA$ is 115 degrees, what is the measure in degrees of angle $A$?

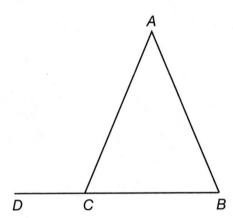

**21.** If $\frac{2w}{3} + 3 = 7$, what is the value of $\frac{3}{2w}$?

**22.** If $\frac{2x+8}{5} = \frac{5x-6}{6}$, what is the value of $x$?

**23.** In the diagram below, sides $OC$ and $OB$ of triangle $OBC$ are congruent. If the measure of angle $OBC$ is 71 degrees, what is the measure in degrees of arc $AD$?

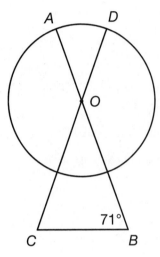

**24.** Evaluate $(\frac{7}{5a^2} + \frac{3}{10a})^a$ for $a = -2$.

**25.** Angle $A$ of right triangle $ABC$ measures 60 degrees and angle $C$ of the triangle measures 30 degrees. If the length of side $BC$ is $16\sqrt{3}$, what is the length of side $AC$?

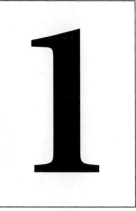

# Algebraic Expressions

## ▶ Variables

A **variable** is a letter or symbol that represents a value. An **algebraic expression** is one or more terms that contain a variable. For instance, $3a$, $y + 1$, and $x^2 + 4x + 4$ are all algebraic expressions. The expression $x^2 + 4x + 4$ is made up of three terms: $x^2$, $4x$, and 4.

## ▶ Parts of a Term

The **coefficient** of a term is the constant in front of the variable. The **base** of a term consists of the variable or variables that follow the coefficient of the term, including the exponents of these variables. The coefficient of the term $4a^3$ is 4, and the base of the term is $a^3$.

## ▶ Like and Unlike Terms

The terms $x$, $3x$, and $8x$ are **like terms**, because these terms all have the same base. The terms $3x$ and $3x^2$ are **unlike terms**, because the variables in each term have different exponents.

## ► Combining Like Terms

Like terms can be combined using the standard operations.

To add like terms, add the coefficients of the terms: $3x + 4x = 7x$, since $3 + 4 = 7$. To subtract like terms, subtract the coefficients of the terms: $9a^2 - 6a^2 = 3a^2$.

To multiply like terms, multiply the coefficients of the terms and add the exponents: $(3b^4)(2b^4) = 6b^8$. To divide like terms, divide the coefficients of the terms and subtract the exponents: $\frac{15x^3}{5x} = 3x^2$.

## ► Combining Unlike Terms

Unlike terms cannot be simplified by adding or subtracting. The expressions $y + 2$, $3a + 3a^2$, $4 - x$, and $6x^6 - 6x$ cannot be simplified any further.

Unlike terms can be simplified by multiplying or dividing. Multiply the coefficients of the terms, and then multiply the variables. Add the exponents of common variables.

What is the product of $6a^2$ and $3a^4$?

The product of the coefficients is $(6)(3) = 18$, and the sum of the exponents of $a$ is $2 + 4 = 6$. The product of $6a^2$ and $3a^4$ is $18a^6$.

What is the product of $4x^2$ and $5xy^4$?

The product of the coefficients is $(4)(5) = 20$. Next, multiply the variables, adding the exponents of common variables: $(x^2)(xy^4) = x^3y^4$. The product of $4x^2$ and $5xy^4$ is $20x^3y^4$.

Unlike terms are divided in a similar way. Divide the coefficients of the terms, and then divide the variables. Subtract the exponents of common variables.

$\frac{25a^3b^5}{5a^2b^2} = 5ab^3$, since $\frac{25}{5} = 5$, the difference of the exponents of the $a$ terms is 1, and the difference of the exponents of the $b$ terms is 3.

## ► Evaluate Expressions Using Substitution

To evaluate an expression given the value of the variable or variables in the expression, substitute the value for the variable in the expression.

What is the value of $7x + 1$ when $x = 5$?
Substitute 5 for $x$ in the expression: $7(5) + 1 = 35 + 1 = 36$.

What is the value of $x^2 + 3x + 5$ when $x = -1$?
The value of $x^2 + 3x + 5$ when $x = -1$ is $(-1)^2 + 3(-1) + 5 = 1 - 3 + 5 = 3$.

# ► Evaluate One Variable in Terms of Another

An equation that contains two variables can be written so that the value of one variable is given in terms of the other. For instance, $y = 3x + 3$ is an equation with two variables, $x$ and $y$, in which the value of one variable, $y$, is written in terms of the other variable, $x$. The value of $y$ is $3x + 3$.

If $4a + 8b = 16$, what is the value of $a$ in terms of $b$?
Isolate $a$ on one side of the equation. Subtract $8b$ from both sides: $4a = 16 - 8b$. Now, divide both sides of the equation by 4: $a = 4 - 2b$. The value of $a$ in terms of $b$ is $4 - 2b$.

*[handwritten: $2(3)^2$ $18-15+3$ $18-18$]*

# ► Practice

**1.** $9a + 12a^2 - 5a =$
   a. $16a$
   b. $16a^2$
   c. $12a^2 - 4a$
   d. $12a^2 + 4a$
   e. $26a^2$

**2.** $\frac{(3a)(4a)}{6(6a^2)} =$
   a. $\frac{1}{3}$
   b. $\frac{2}{a}$
   c. $\frac{1}{3a}$
   d. $\frac{1}{3a^2}$
   e. $2$

*[handwritten: $\frac{12a^2}{6(6a^2)}$  $\frac{12a^2}{36a^2}$  $\frac{1}{3}$]*

**3.** $\frac{(5a + 7b)b}{(b + 2b)} =$
   a. $4a$
   b. $4ab$
   c. $2a + 4b$
   d. $\frac{5a + 7b}{3}$
   e. $5a + 5b$

*[handwritten: $\frac{5ab + 7b^2}{b + 2b}$  $\frac{5ab + 7b^2}{3b}$  $\frac{5ab + 7b}{3}$]*

*[handwritten: $8x^2y^2 + 6x^2y^2$]*

**4.** $(2x^2)(4y^2) + 6x^2y^2 =$
   a. $12x^2y^2$
   b. $14x^2y^2$
   c. $2x^2 + 4y^2 + 6x^2y^2$
   d. $8x^2 + y^2 + 6x^2y^2$
   e. $8x^4y^4 + 6x^2y^2$

**5.** When $x = 3$, $2x^2 - 5x + 3 =$
   a. $-6$
   b. $0$
   c. $6$
   d. $33$
   e. $36$

*[handwritten: $6^2 - 15+3$  $6^2 - 28$  $36 - 28$]*

**6.** When $a = -2$, $\frac{7a}{a^2 + a} =$
   a. $-14$
   b. $-7$
   c. $-\frac{7}{4}$
   d. $\frac{7}{4}$
   e. $7$

*[handwritten: $\frac{-14}{4-2}$  $\frac{-14}{2}$  $-7$]*

**7.** When $x = -2$, $\frac{y^2}{x^2} + \frac{y}{-2x} =$
   a. $\frac{3y}{4}$
   b. $\frac{y^3}{4}$
   c. $\frac{(-y^2 + y)}{4}$
   d. $\frac{(y^2 - y)}{4}$
   e. $\frac{(y^2 + y)}{4}$

*[handwritten: $\frac{y^2}{4} + \frac{y}{4}$  $\frac{(y^2 + y)}{4}$]*

**8.** When $p = 6$, $\frac{4p(p + r)}{pr} =$
   a. $8$
   b. $24 + r$
   c. $24 + 3r$
   d. $\frac{6}{r}$
   e. $\frac{24 + 4r}{r}$

*[handwritten: $4p^2 + 4pr$  $\frac{24(6 + r)}{6r}$  $\frac{144 + 24y}{6}$  $\frac{24 + 4r}{r}$]*

**9.** When $a = 3$, $(4a^2)(3b^3 + a) - b^3 =$
   **a.** $72b^3 + 108$
   **b.** $107b^3 + 39$
   **c.** $107b^3 + 108$
   **d.** $108b^3 + 39$
   **e.** $216$

$(4 \cdot 9)(3b^3 + 3) - b^3$
$(36)(3b^2 + 3)$
$108b^2 + 108 - b^3$
$107b^3 + 108$

**10.** When $c = 1$ and $d = 4$, $\frac{(cd)^2}{c+d} =$
   **a.** $\frac{16}{5}$
   **b.** $4$
   **c.** $\frac{17}{4}$
   **d.** $5$
   **e.** $\frac{25}{4}$

$\frac{(1 \cdot 4)(1 \cdot 4)16}{1 + 4} \quad \frac{16}{5}$

$\begin{array}{ccc} 1 & 4 & 16 \end{array}$

**11.** When $x = 2$ and $y = 3$, $\frac{6x^2}{2y^2} + \frac{4x}{3y} =$
   **a.** $\frac{4}{9}$
   **b.** $\frac{4}{3}$
   **c.** $\frac{20}{9}$
   **d.** $\frac{21}{9}$
   **e.** $\frac{13}{3}$

$\frac{6x}{2y} + \frac{4x}{3y}$
$\frac{6(2) + 4(2)}{2(3) + 3(3)}$
$\frac{12 + 8}{6 + 9} = \frac{20}{15}$

**12.** When $a = 1$ and $b = -1$, $ab + \frac{a}{b} + a^2 - b^2 =$
   **a.** $-4$
   **b.** $-3$
   **c.** $-2$
   **d.** $-1$
   **e.** $0$

$(1 \cdot -1) + \frac{1}{-1} + 1 - 1$
$-1 + -1 + 1 - 1$

**13.** If $\frac{3}{2}g = 9h - 15$, what is the value of $g$ in terms of $h$?
   **a.** $\frac{1}{6}h + \frac{5}{3}$
   **b.** $6h - 15$
   **c.** $6h - 10$
   **d.** $\frac{27}{2}h - \frac{45}{2}$
   **e.** $18h - 30$

$g = 6h + 10$

**14.** If $7a + 20b = 28 - b$, what is the value of $a$ in terms of $b$?
   **a.** $4 - 3b$
   **b.** $4 + 3b$
   **c.** $4 - \frac{19}{7}b$
   **d.** $4 - \frac{22}{7}b$
   **e.** $-\frac{b+4}{3}$

$7a + 20b = 28 - b$
$7a + 21b = 28$
$7a = 28 - 21b$
$a = 4 - 3b$

**15.** If $4(\frac{x}{y} + 1) = 10$, what is the value of $y$ in terms of $x$?
   **a.** $-\frac{1}{4}x$
   **b.** $\frac{1}{4}x$
   **c.** $\frac{4}{9}x$
   **d.** $\frac{2}{3}x$
   **e.** $\frac{3}{2}x$

$4(\frac{x}{y} + 1) = 10$
$4\frac{x}{y} + 4 = 10$
$\frac{4x}{y} + 4 = 10 \quad \frac{4x}{y} = 6$

**16.** If $fg + 2f - g = 2 - (f + g)$, what is the value of $g$ in terms of $f$?
   **a.** $-1$
   **b.** $\frac{1}{f}$
   **c.** $\frac{4}{f}$
   **d.** $2 - 2f$
   **e.** $\frac{2 - 3f}{f}$

$\frac{fg + 2f - g}{g} = \frac{2 - (f + g)}{g}$
$f + \frac{2f}{g} = 2 - \frac{fg}{g} f$

**17.** If $a(3a) - b(4 + a) = -(a^2 + ab)$, what is the value of $b$ in terms of $a$?
   **a.** $\frac{1}{2}a$
   **b.** $\frac{1}{2}a^2$
   **c.** $2a$
   **d.** $a^2$
   **e.** $4a^2$

$a(3a) - b4 - ba = -1(a^2 + ab)$
$3a^2 - b4 - ba = -a^2 - ba$
$\sqrt{3a^2 - b4} = a^2$
$3a - 2b = a$
$-2b = -2a$
$b = a$

**18.** If $4g^2 - 1 = 16h^2 - 1$, what is the value of $g$ in terms of $h$?

$\sqrt{4g^2} = \sqrt{1 + 16h^2}$

a. $h$

$4g = 1 + 16h$

b. $2h$

c. $4h$

$g = \frac{1}{4} + 4h$

d. $h^2$

e. $4h^2$

**19.** If $8x^2 - 4y^2 + x^2 = 0$, what is the value of $x$ in terms of $y$?

$8x^2 - 4y^2 + x^2 = 0$

a. $\frac{2}{3}y$

$9x^2 - 4y^2 = 0$

b. $\frac{3}{2}y$

$\sqrt{9x^2} = \sqrt{4y^2}$

c. $\frac{4}{9}y^2$

$3x = 2y$

d. $\frac{2}{3}y^2$

e. $\frac{3}{2}y^2$

**20.** If $\frac{10(x^2y)}{xy^2} = 5y$, what is the value of $y$ in terms of $x$?

a. $\sqrt{x}$

$\frac{10(x^2y)}{xy^2} \cdot 10\frac{x}{y} = 5y$

b. $\sqrt{2x}$

c. $2\sqrt{x}$

$10x = 5y^2$

d. $2x$

$x = \frac{y^2}{2}$

e. $\frac{x^2}{2}$

$\frac{12}{6} \quad \frac{18}{9} \quad \frac{8}{9}$

$\frac{6 \cdot 4}{2 \cdot 9} \quad \frac{4 \cdot 2}{3 \cdot 3}$

$2\sqrt{\frac{24}{18}} + \frac{8}{9}$

$\frac{12}{9} + \frac{8}{9} = \frac{20}{9}$

$fg + 2f - g = 2 - 1(f + g)$

$fg + 2f + g = 2 - f - g$

$fg + 2f = 2 - f$

$fg = 2 - 3f$

$g = \frac{2 - 3f}{f}$

$f + \frac{2f}{g} = 2 - f$

$2f + \frac{2f}{g} = 2 - f$

$2f = 2 - \frac{2f}{g}$

$f = 1 - \frac{f}{g}$

# 2 ▶ Solving Equations and Inequalities

**A**n **equation** is an expression that contains an equals sign; $3x + 2 = 8$ is an equation. An inequality is an expression that contains one of the following symbols: $<$, $>$, $\leq$, or $\geq$. Both sides of an equation are equal. One side of an inequality may be less than, less than or equal to, greater than, or greater than or equal to the other side of the inequality.

## ▶ Basic Equations and Inequalities

Basic equations and inequalities can be solved with one step. To solve an equation, isolate the variable on one side of the equals sign by adding, subtracting, multiplying, or dividing. To solve an inequality, isolate the variable on one side of the inequality symbol using the same operations.

To solve the equation $x - 2 = 4$, add 2 to both sides of the equation: $x - 2 + 2 = 4 + 2, x = 6$.

To solve the equation $x + 5 = 6$, subtract 5 from both sides of the equation: $x + 5 - 5 = 6 - 5, x = 1$.

To solve the inequality $\frac{x}{4} < 8$, multiply both sides of the equation by 4: $(4)(\frac{x}{4}) < (4)(8), x < 32$.

To solve the inequality $3x \geq 6$, divide both sides of the inequality by 3: $\frac{3x}{3} \geq \frac{6}{3}, x \geq 2$.

If solving an inequality requires multiplying or dividing both sides of the inequality by a negative number, you must reverse the inequality symbol.

Solve the inequality $-2x < 10$ for $x$.

Divide both sides of the inequality by $-2$: $\frac{-2x}{-2} > \frac{10}{-2}, x > -5$. Notice that the inequality symbol has changed direction. Values of $x$ that are less than $-5$, if substituted into $-2x < 10$, would make the left side of the inequality greater than the right side of the inequality. That is why you must reverse the inequality symbol: $x > -5$. This is the solution set for $-2x < 10$. Only reverse the inequality symbol when multiplying or dividing both sides of the inequality by a negative number.

## ▶ Multi-Step Equations and Inequalities

Multi-step equations and inequalities require two or more steps to solve. Use combinations of the operations above to isolate the variable on one side of the equals sign or inequality symbol. You may also have to raise both sides of the equation to an exponent, or take a root of both sides of the equation.

To solve the inequality $10x - 2 > 18$, add 2 to both sides of the inequality: $10x - 2 + 2 > 18 + 2, 10x > 20$. Then, divide both sides of the inequality by 10: $\frac{10x}{10} > \frac{20}{10}, x > 2$.

To solve the equation $x^2 + 4 = 20$, subtract 4 from both sides of the equation: $x^2 + 4 - 4 = 20 - 4, x^2 = 16$. Both 4 and $-4$ square to 16, so the solutions to $x^2 + 4 = 20$ are $x = 4$ and $x = -4$.

## ▶ Cross Multiplying

If one or both sides of an equation or inequality contain fractions, begin solving the equation or inequality by cross multiplying. Multiply the numerator of the fraction on the left side of the equation or inequality by the denominator of the fraction on the right side of the equation or inequality. Then, multiply the numerator of the fraction on the right side of the equation or inequality by the denominator of the fraction on the left side of the equation or inequality, and set the two products equal to each other.

If $\frac{3x}{2} = \frac{18x - 6}{6}$, what is the value of $x$?

Begin by cross multiplying: $(3x)(6) = (2)(18x - 6), 18x = 36x - 12$. Now, solve the equation by adding and dividing:

$18x + 12 = 36x - 12 + 12$

$18x + 12 = 36x$

$18x - 18x + 12 = 36x - 18x$

$12 = 18x$

$\frac{12}{18} = \frac{18x}{18}$

$x = \frac{12}{18} = \frac{2}{3}$

## ▶ Undefined Expressions

A fraction is **undefined** when its denominator is equal to zero. To find when an algebraic fraction is undefined, set the denominator of the fraction equal to zero and solve the equation for the value of the variable.

For what value of $x$ is the fraction $\frac{1}{2x+4}$ undefined?

Set the denominator of the fraction, $2x + 4$, equal to 0 and solve for $x$:

$2x + 4 = 0$

$2x + 4 - 4 = 0 - 4$

$2x = -4$

$\frac{2x}{2} = \frac{-4}{2}$

$x = -2$. The fraction $\frac{1}{2x+4}$ is undefined when $x = -2$.

## ▶ Forming Algebraic Equations from Word Problems

The SAT will often describe a situation that must be transformed into an algebraic equation in order to be solved. Read carefully for key words. Phrases such as *less than* signal subtraction; *greater than* signals addition. Words such as *twice* signal multiplication, while words such as *half* signal division.

If two less than three times a number is two more than twice the number, what is the number?

Transform the situation into an algebraic expression. If $x$ is the number, then *two less than three times a number* is $3x - 2$. Twice the number is $2x$. Two more than that is $2x + 2$. The key word *is* means *equals*. The phrase *two less than three times a number is two more than twice the number* can be represented algebraically as $3x - 2 = 2x + 2$. Now, solve the equation for $x$:

$3x - 2 = 2x + 2$

$3x - 2x - 2 = 2x - 2x + 2$

$x - 2 = 2$

$x - 2 + 2 = 2 + 2$

$x = 4$

## ▶ Practice

**1.** If $a - 12 = 12$, $a =$
   a. $-12$   $a = 24$
   b. 0
   c. 1
   d. 12
   e. 24

**2.** If $6p \geq 10$, then
   a. $p \geq \frac{3}{5}$
   b. $p \leq \frac{3}{5}$
   c. $p \geq \frac{5}{3}$
   d. $p \leq \frac{5}{3}$
   e. $p \geq 2$

**3.** If $x + 10 = 5$, $x =$
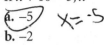
   a. $-5$   $x = -5$
   b. $-2$
   c. $\frac{1}{2}$
   d. $2$
   e. $5$

**4.** If $\frac{k}{8} = 8$, $k =$
   a. $\frac{1}{8}$
   b. $\sqrt{8}$

   c. $8$
   d. $16$
   e. $64$

**5.** The inequality $-3n < 12$ is equivalent to
   a. $n < 4$
   b. $n < -4$        $\frac{3}{n} > 4$
   c. $n > 4$
   d. $n > -4$
   e. $n > 36$

**6.** If $9a + 5 = -22$, what is the value of $a$?
   a. $-27$        $9a = -29$
   b. $-9$         $a = -3$
   c. $-3$
   d. $-2$
   e. $-\frac{17}{9}$

**7.** The inequality $3x - 6 \le 4(x + 2)$ is equivalent to
   a. $x \ge -14$   $3x - 6 \le 4x + 8$
   b. $x \le -14$   $3x \le 4x + 14$
   c. $x \le -8$    $-x \le 14$
   d. $x \ge -8$
   e. $x \ge 2$     $x \ge 14$

**8.** If $6x - 4x + 9 = 6x + 4 - 9$, what is the value of $x$?
   a. $-9$        $-4x + 9 = 4 - 9$
   b. $-\frac{7}{2}$     $9$
   c. $-1$        $-4x + 18 = 4$
   d. $\frac{7}{2}$      $-4x = -14$
   e. $9$         $x = \frac{14}{-4} - \frac{7}{2}$

**9.** What is the solution set for the inequality $-8(x + 3) \le 2(-2x + 10)$?
   a. $x \ge -11$   $-8x - 24 \le -4x + 20$
   b. $x \le -11$
   c. $x = 11$      $-4x = 44$
   d. $x \ge 11$
   e. $x \le 11$    $x \ge 11$

**10.** If $\frac{3c^2}{6c} + 9 = 15$, what is the value of $c$?
   a. $\frac{1}{2}$      $\frac{5c^2}{6c} + 9 = 15$
   b. $2$
   c. $3$         $6c$
   d. $6$         $\frac{5c^2}{6c} = 6$
   e. $12$

**11.** If $\frac{w}{w+8} = -\frac{6}{18}$, what is the value of $w$?
   a. $-6$        $-8w = 6w + 48$
   b. $-2$
   c. $-1$        $-18w = 48$
   d. $4$
   e. $10$        $v = -2$

**12.** If $\frac{10x}{7} = \frac{5x - 10}{3}$, what is the value of $x$?
   a. $\frac{7}{5}$
   b. $3$         $30x = 35x - 70$
   c. $5$
   d. $14$        $-5x = -70$
   e. $30$        $x = 14$

**13.** If $\frac{4a+4}{7} = -\frac{2-3a}{4}$, what is the value of $a$?

   **a.** $-\frac{30}{37}$

   **b.** $\frac{12}{5}$

   **c.** 4

   **d.** 6

   **e.** 16

*[handwritten: $16a-16 = -21a$; $-16 = 14+5a$; $-16a-16 = -14+21a$; $-16a = 2+21a$; $5a=2$]*

**14.** If $\frac{6}{-y-1} = \frac{10}{-2y-3}$, what is the value of $y$?

   **a.** $-4$

   **b.** $-2$

   **c.** 2

   **d.** 4 *(circled)*

   **e.** 14

*[handwritten: $-10y-10 = -12y-18$; $-10y+8 = -12y$; $8=2y$; $y=4$]*

**15.** If $\frac{5g}{g} = \frac{g+7}{g-1}$, what is the value of $g$?

   **a.** $\frac{1}{2}$

   **b.** 2

   **c.** 3 *(circled)*

   **d.** 4

   **e.** 5

*[handwritten: $5g^2-5g = g^2+7g$; $5g^2 = g^2+12g$; $4g^2 = 12$; $g^2 = 3$]*

**16.** If $\frac{1}{2}x + 6 = -x - 3$, what is the value of $-2x$?

   **a.** $-12$ *(circled)*

   **b.** $-6$

   **c.** $-3$

   **d.** 6

   **e.** 12

*[handwritten: $1\frac{1}{2}x+6 = -3$; $1\frac{1}{2}x = -9$; $2x = 12$]*

**17.** If $3x - 8 = \frac{9x+5}{2}$, what is the value of $x + 7$?

   **a.** $-7$

   **b.** 0

   **c.** 7

   **d.** 10

   **e.** 14

*[handwritten: $6x-16 = 9x+5$; $6x = 9x+21$; $-3x = 21$; $-3x+21=0$; $3x+21=0$]*

**18.** If $9x + \frac{8}{3} = \frac{8}{3}x + 9$, what is the value of $\frac{3}{8}x$?

   **a.** $\frac{3}{8}$

   **b.** 1

   **c.** $\frac{8}{3}$

   **d.** $\frac{3}{8}$

   **e.** 9

**19.** If $8x + 4 = 14$, what is the value of $4x + 2$?

   **a.** $\frac{5}{8}$

   **b.** $\frac{5}{4}$

   **c.** $\frac{5}{2}$

   **d.** 7 *(circled)*

   **e.** 28

*[handwritten: $4x+2 = 7$]*

**20.** If $11c - 7 = 8$, what is the value of $33c - 21$?

   **a.** $\frac{15}{11}$

   **b.** $\frac{8}{3}$

   **c.** 16

   **d.** 24

   **e.** 45

*[handwritten: $33c-21 = 24$]*

**21.** For what value of $x$ is the fraction $\frac{-8}{x+8}$ undefined?

   **a.** $-8$

   **b.** $-1$

   **c.** $\frac{1}{2}$

   **d.** 1

   **e.** 8

**22.** For what value of $d$ is the fraction $\frac{3d}{6d}$ undefined?

   **a.** $-\frac{1}{3}$

   **b.** $-\frac{1}{6}$

   **c.** 0

   **d.** $\frac{1}{3}$ *(circled)*

   **e.** $\frac{1}{6}$

**23.** For what value of $a$ is the fraction $\frac{2a-18}{6a+18-4a}$ undefined?
   **a.** −18
   **b.** −9
   **c.** −1
   **d.** 1
   **e.** 9

**24.** For what value of $y$ is the fraction $\frac{3y-6}{8-8y+4}$ undefined?
   **a.** −2
   **b.** $-\frac{1}{2}$
   **c.** $\frac{3}{2}$
   **d.** 2
   **e.** 4

**25.** For what values of $x$ is the fraction $\frac{2x+10}{2x(x-5)}$ undefined?
   **a.** 0, −5
   **b.** 0, 5
   **c.** 5, 10
   **d.** −5, 10
   **e.** 0, 5, −5

**26.** If five less than three times a number is equal to 10, what is the value of that number?
   **a.** $\frac{10}{3}$
   **b.** 5
   **c.** $\frac{25}{3}$
   **d.** 15
   **e.** 25

**27.** If three more than one-fourth of a number is three less than the number, what is the value of the number?
   **a.** $\frac{3}{4}$
   **b.** 4
   **c.** 6
   **d.** 8
   **e.** 12

**28.** The sum of three consecutive integers is 63. What is the value of the largest integer?
   **a.** 18
   **b.** 19
   **c.** 20
   **d.** 21
   **e.** 22

**29.** The sum of four consecutive, odd whole numbers is 48. What is the value of the smallest number?
   **a.** 9
   **b.** 11
   **c.** 13
   **d.** 15
   **e.** 17

**30.** The sum of three consecutive, even integers is −18. What is the value of the smallest integer?
   **a.** −12
   **b.** −10
   **c.** −8
   **d.** −6

   **e.** −4

# 3 ▶ Quadratic Expressions and Equations

**A** quadratic expression is an expression that contains an $x^2$ term. The expressions $x^2 - 4$ and $x^2 + 3x + 2$ are two examples of quadratic expressions. A **quadratic equation** is a quadratic expression set equal to a value. The equation $x^2 + 3x + 2 = 0$ is a quadratic equation.

## ▶ Multiplying Binomials

A **binomial** is an expression with two terms, each with a different base; $(x - 2)$ and $(5x + 1)$ are both binomials.

To multiply binomials, you must multiply each term by every other term and add the products. The acronym **FOIL** can help you to remember how to multiply binomials. FOIL stands for First, Outside, Inside, and Last.

When multiplying two binomials, for example, $(x + 1)$ and $(x + 2)$, begin by multiplying the **first** term in each binomial: $(\mathbf{x} + 1)(\mathbf{x} + 2)$, $(x)(x) = x^2$.

Next, multiply the **outside** terms: $(\mathbf{x} + 1)(x + \mathbf{2})$, $(x)(2) = 2x$.

Then, multiply the **inside** terms: $(x + \mathbf{1})(\mathbf{x} + 2)$, $(1)(x) = x$.

Finally, multiply the **last** terms: $(x + \mathbf{1})(x + \mathbf{2})$, $(1)(2) = 2$.

To find the product of $(x + 1)(x + 2)$, add the four products: $x^2 + 2x + x + 2 = x^2 + 3x + 2$.

# ▶ Factoring Quadratic Expressions

Multiplying the binomials $(x + 1)$ and $(x + 2)$ creates the quadratic expression $x^2 + 3x + 2$. That expression can be broken back down into $(x + 1)(x + 2)$ by factoring.

A quadratic trinomial (a **trinomial** is an expression with three terms) that begins with the term $x^2$ can be factored into $(x + a)(x + b)$. Factoring is the reverse of FOIL. Find two numbers, $a$ and $b$, that multiply to the third value of the trinomial (the constant) and that add to the coefficient of the second value of the trinomial (the $x$ term).

Given the quadratic $x^2 + 6x + 8$, you can find its factors by finding two numbers whose product is 8 and whose sum is 6. The numbers 1 and 8, and, 4 and 2 multiply to 8, but only 4 and 2 add to 6. The factors of $x^2 + 6x + 8$ are $(x + 2)$ and $(x + 4)$. You can check your factoring by using FOIL: $(x + 2)(x + 4) = x^2 + 4x + 2x + 8 = x^2 + 6x + 8$.

What are the factors of $2x^2 + 9x + 9$?

This quadratic will be factored into $(2x + a)(x + b)$. Find two numbers that multiply to 9. Two times one of those numbers plus the other must equal 9, the coefficient of the second term of the quadratic trinomial. The numbers 1 and 9, and the numbers 3 and 3 multiply to 9. Two times 3 plus 3 is equal to 9, so the factors of $2x^2 + 9x + 9$ are $(2x + 3)$ and $(x + 3)$.

# ▶ Solving Quadratic Equations

Quadratic equations have two solutions. To solve a quadratic equation, combine like terms and place all terms on one side of the equals sign, so that the quadratic is equal to 0. Then, factor the quadratic and find the values of $x$ that make each factor equal to 0. The values that solve a quadratic are the **roots** of the quadratic.

For what values of $x$ does $x^2 - 5x = -6$?

First, combine like terms and place all terms on one side of the equals sign. Add 6 to both sides of the equation; $x^2 - 5x + 6 = 0$. Now, factor the quadratic. Find two numbers that multiply to 6 and add to −5. The numbers 1 and 6, −1 and −6, 3 and 2, and −3 and −2 multiply to 6. Only −3 and −2 add to −5. The factors of $x^2 - 5x + 6$ are $(x - 3)$ and $(x - 2)$. If either of these factors equals 0, the expression $x^2 - 5x + 6$ equals 0. Set each factor equal to 0, and solve for $x$; $x - 3 = 0$, $x = 3$; $x - 2 = 0$, $x = 2$. The values 2 and 3 for $x$ make $x^2 - 5x + 6$ equal to 0; therefore, these values make $x^2 - 5x$ equal to −6.

## ▶ Undefined Expressions

As you saw in the last chapter, a fraction is undefined when its denominator is equal to zero. If the denominator of a fraction is a quadratic, factor the quadratic and set the factors equal to 0. The values that make the quadratic equal to 0 are the values that make the fraction undefined.

For what values of $x$ is the fraction $\frac{2x}{x^2 + 7x + 10}$ undefined?

Find the roots of $x^2 + 7x + 10$. The numbers 1 and 10, –1 and –10, 5 and 2, and –5 and –2 multiply to 10, but only 5 and 2 add to 7. The factors of $x^2 + 7x + 10$ are $(x + 5)(x + 2)$. Since $x + 5 = 0$ when $x = -5$ and $x + 2 = 0$ when $x = -2$, the values of $x$ that make the fraction undefined are –5 and –2.

## ▶ Graphs of Quadratic Equations

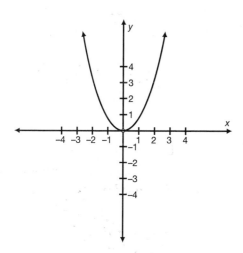

The graph of the equation $y = x^2$, shown at left, is a parabola. Since the $x$ value is squared, the positive values of $x$ yield the same $y$ values as the negative values of $x$. The graph of $y = x^2$ has its vertex at the point (0,0). The vertex of a parabola is the turning point of the parabola. It is either the minimum or maximum $y$ value of the graph. The graph of $y = x^2$ has its minimum at (0,0). There are no $y$ values less than 0 on the graph.

The graph of $y = x^2$ can be translated around the coordinate plane. While the parabola $y = x^2$ has its vertex at (0,0), the parabola $y = x^2 - 1$ has its vertex at (0,–1). After the $x$ term is squared, the graph is shifted down one unit. A parabola of the form $y = x^2 - c$ has its vertex at (0,–c) and a parabola of the form $y = x^2 + c$ has its vertex at (0,c).

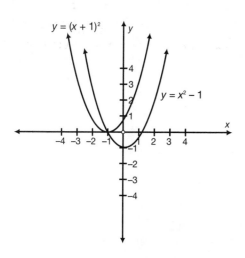

The parabola $y = (x + 1)^2$ has its vertex at (–1,0). The $x$ value is increased before it is squared. The minimum value of the parabola is when $y = 0$ (since $y = (x + 1)^2$ can never have a negative value). The expression $(x + 1)^2$ is equal to 0 when $x = -1$. A parabola of the form $y = (x - c)^2$ has its vertex at (c,0) and a parabola of the form $y = (x + c)^2$ has its vertex at (0,–c).

What are the coordinates of the vertex of the parabola formed by the equation $y = (x - 2)^2 + 3$?

To find the $x$ value of the vertex, set $(x - 2)$ equal to 0; $x - 2 = 0$, $x = 2$. The $y$ value of the vertex of the parabola is equal to the constant that is added to or subtracted from the $x$ squared term. The $y$ value of the vertex is 3, making the coordinates of the vertex of the parabola (2,3).

► **Practice**

*(handwritten: $x^2 + 4x - 21$)*
*(handwritten: $x^2 + 7x - 3x - 21$)*

**1.** What is the product of $(x - 3)(x + 7)$?
   a. $x^2 - 21$
   b. $x^2 - 3x - 21$
   c. $x^2 + 4x - 21$
   d. $x^2 + 7x - 21$
   e. $x^2 - 21x - 21$

*(handwritten: $x^2 + 7 - 3x + 21$)*
*(handwritten: $x^2 + 4x - 21$)*

**2.** What is the product of $(x - 6)(x - 6)$?
   a. $x^2 + 36$
   b. $x^2 - 36$
   c. $x^2 - 12x - 36$
   d. $x^2 - 12x + 36$
   e. $x^2 - 36x + 36$

*(handwritten: $x^2 - 6x - 6x - 36$)*
*(handwritten: $x^2 - 12x - 36$)*

**3.** What is the product of $(x - 1)(x + 1)$?
   a. $x^2 - 1$
   b. $x^2 + 1$
   c. $x^2 - x - 1$
   d. $x^2 - x + 1$
   e. $x^2 - 2x - 1$

*(handwritten: $x^2 + x - x - 1$)*
*(handwritten: $x^2 - 1$)*

**4.** What is the value of $(x + c)^2$?
   a. $x^2 + c^2$
   b. $x^2 + cx + c^2$
   c. $x^2 + c^2x^2 + c^2$
   d. $x^2 + cx + c^2x + c^2$
   e. $x^2 + 2cx + c^2$

**5.** What is the product of $(2x + 6)(3x - 9)$?
   a. $5x^2 - 54$
   b. $6x^2 - 54$
   c. $6x^2 + 18x - 15$
   d. $6x^2 - 18x - 15$
   e. $6x^2 + 36x - 54$

*(handwritten: $6x^2 - 18x + 18x - 54$)*
*(handwritten: $6x^2 - 54$)*

**6.** What are the factors of $x^2 - x - 6$?
   a. $(x - 3)(x - 2)$
   b. $(x - 3)(x + 2)$
   c. $(x + 3)(x - 2)$
   d. $(x - 6)(x + 1)$
   e. $(x - 1)(x + 6)$

**7.** What is one factor of $x^2 - 4$?
   a. $x^2$
   b. $-4$
   c. $(x - 1)$
   d. $(x + 2)$
   e. $(x - 4)$

*(handwritten: $(x + 2)$)*

**8.** What are the factors of $x^2 - 11x + 28$?
   a. $(x - 4)(x - 7)$
   b. $(x - 4)(x + 7)$
   c. $(x + 14)(x - 2)$
   d. $(x - 14)(x + 2)$
   e. $(x - 11)(x + 28)$

*(handwritten: $(x + ?)(x - 11 - 7)$, $28$, $4$)*

**9.** What are the roots of $x^2 - 18x + 32 = 0$?
   a. $-4$ and $-8$
   b. $-4$ and $8$
   c. $-2$ and $-16$
   d. $2$ and $-16$
   e. $2$ and $16$

*(handwritten: $(x - 16)(x - ?)$, $32$, $x = 16, 2$)*

**10.** What are the roots of $x^2 + 8x - 48 = 0$?
   a. $-6$ and $-8$
   b. $-6$ and $8$
   c. $4$ and $-12$
   d. $-4$ and $12$
   e. $-4$ and $-12$

*(handwritten: $-48$, $-4$, $12$, $8x$)*

**11.** The fraction $\frac{(x+9)}{(x^2-81)}$ is equivalent to

a. $\frac{1}{(x-9)}$

b. $\frac{1}{(x+9)}$

c. $\frac{1}{(x^2-81)}$

d. $\frac{(x+9)}{(x-9)}$

e. $\frac{(x-9)}{(x+9)}$

**12.** The fraction $\frac{(x^2-6x-16)}{(x^2-x-56)}$ is equivalent to

a. $\frac{(x+2)}{(x-8)}$

b. $\frac{(x+2)}{(x+7)}$

c. $\frac{(x+7)}{(x+2)}$

d. $\frac{(x-4)}{(x-8)}$

e. $\frac{(x-8)}{(x-8)}$

**13.** The fraction $\frac{(x^2-4x-45)}{(x^2+11x+30)}$ is equivalent to

a. $\frac{(x+5)}{(x-9)}$

b. $\frac{(x+5)}{(x+6)}$

c. $\frac{(x-9)}{(x+6)}$

d. $\frac{(x-9)}{(x+5)}$

e. $\frac{(x+5)}{(x+5)}$

**14.** The fraction $\frac{(x^2+14x+33)}{(x^2+8x-33)}$ is equivalent to

a. $-1$

b. $\frac{(x+3)}{(x-3)}$

c. $\frac{(x-3)}{(x+3)}$

d. $\frac{(x+3)}{(x+11)}$

e. $\frac{(x-3)}{(x+11)}$

**15.** If $x^2 - x = 12$, what is one possible value of $x$?

a. $-6$

b. $-2$

c. $1$

d. $3$

e. $4$

**16.** If $x^2 - 3x - 30 = 10$, what is one possible value of $x$?

a. $-10$

b. $3$

c. $5$

d. $8$

e. $10$

**17.** If $\frac{x+5}{4} = \frac{6x-6}{x+4}$, what is one possible value of $x$?

a. $-5$

b. $-4$

c. $3$

d. $5$

e. $11$

**18.** If $(x-2)(x+6) = -16$, what is the value of $x$?

a. $-6$

b. $-4$

c. $-2$

d. $6$

e. $7$

**19.** If $(x-7)(x-5) = -1$, what is the value of $x$?

   **a.** −6

   **b.** −4

   **c.** −2

   **d.** 6

   **e.** 7

**20.** The square of a number is equal to two less than three times the number. What are two possible values of the number?

   **a.** 1, 2

   **b.** −1, 2

   **c.** 1, −2

   **d.** −1, −2

   **e.** 2, 3

**21.** What is one value that makes the fraction $\frac{(4x-8)}{(x^2+x-42)}$ undefined?

   **a.** −7

   **b.** −6

   **c.** −2

   **d.** 2

   **e.** 21

**22.** What values make the fraction $\frac{(x^2+8x+7)}{(x^2-8x+7)}$ undefined?

   **a.** −1, −7

   **b.** 1, −7

   **c.** −1, 7

   **d.** 1, 7

   **e.** −1, −7, 1, 7

**23.** What is one value that makes the fraction $\frac{(x-16)}{(x^2-16)}$ undefined?

   **a.** −16

   **b.** −4

   **c.** −1

   **d.** 1

   **e.** 16

**24.** What is one value that makes the fraction $\frac{(x^2)}{(9x^2-1)}$ undefined?

   **a.** −3

   **b.** 0

   **c.** $\frac{1}{3}$

   **d.** 3

   **e.** 9

**25.** What values make the fraction $\frac{(x^2-36)}{(2x^2-25x+72)}$ undefined?

   **a.** −8, −9

   **b.** −6, 6

   **c.** $\frac{9}{2}$, 4

   **d.** $\frac{9}{2}$, 8

   **e.** 8, 9

**26.** What is the vertex of the parabola whose equation is $y = x^2 + 4$?

   **a.** (−4,0)

   **b.** (0,−4)

   **c.** (−2,0)

   **d.** (0,4)

   **e.** (2,8)

**27.** Which of the following is the equation of a parabola whose vertex is $(-3,-4)$?

**a.** $y = x^2 - 7$
**b.** $y = (x - 3)^2 - 4$
**c.** $y = (x - 4)^2 - 3$
**d.** $y = (x + 4)^2 - 5$
**e.** $y = (x + 3)^2 - 4$

**28.** What is the vertex of the parabola whose equation is $y = (x + 2)^2 + 2$?

**a.** $(2,2)$
**b.** $(-2,2)$
**c.** $(2,-2)$
**d.** $(-2,-2)$
**e.** $(2,18)$

**29.** Which of the following is the equation of a parabola whose vertex is $(5,0)$?

**a.** $y = x^2 - 5$
**b.** $y = x^2 + 5$
**c.** $y = (x - 5)^2$
**d.** $y = (x + 5)^2$
**e.** $y = x^2 - 25$

**30.** What is the equation of the parabola shown below?

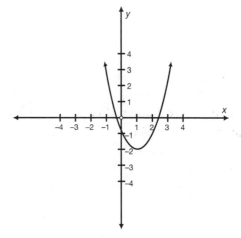

**a.** $y = (x - 1)^2 - 2$
**b.** $y = (x + 1)^2 - 2$
**c.** $y = (x - 1)^2 + 2$
**d.** $y = (x + 1)^2 + 2$
**e.** $y = x^2 - 3$

# Factoring and Multiplying Polynomials

 **polynomial** is an expression with more than one term, each with a different base. A binomial is a polynomial with two terms; a trinomial is a polynomial with three terms.

## ▶ Multiplying Polynomials

To multiply two polynomials, multiply every term of the first polynomial by every term of the second polynomial. Then, add the products and combine like terms.

What is $(x + 4)(x^2 + 6x + 8)$?
$(\mathbf{x} + 4)(\mathbf{x^2} + 6x + 8)$, $(x)(x^2) = x^3$
$(\mathbf{x} + 4)(x^2 + \mathbf{6x} + 8)$, $(x)(6x) = 6x^2$
$(\mathbf{x} + 4)(x^2 + 6x + \mathbf{8})$, $(x)(8) = 8x$
$(x + \mathbf{4})(\mathbf{x^2} + 6x + 8)$, $(4)(x^2) = 4x^2$
$(x + \mathbf{4})(x^2 + \mathbf{6x} + 8)$, $(4)(6x) = 24x$
$(x + \mathbf{4})(x^2 + 6x + \mathbf{8})$, $(4)(8) = 32$

Add the products and combine like terms: $x^3 + 6x^2 + 8x + 4x^2 + 24x + 32 = x^3 + 10x^2 + 32x + 32$.

## ▶ Factoring Polynomials

To factor the kinds of polynomials you will encounter on the SAT, first look for a factor common to every term in the polynomial, such as $x$ or a constant. After factoring out that value, factor the monomial, binomial, or quadratic that remains.

What are the factors of $3x^2 + 15x - 108$?

First, factor out the constant 3: $3(x^2 + 5x - 36)$. Then, factor the quadratic that remains: $x^2 + 5x - 36 = (x - 4)(x + 9)$, and $3x^2 + 15x - 108 = 3(x - 4)(x + 9)$.

## ▶ Finding Roots

As you saw in the last chapter, the roots of an equation are the values that make the equation true.

What are the roots of $x^3 - 9x^2 - 10x = 0$?

First, factor out the variable $x$: $x(x^2 - 9x - 10)$. Then, factor the quadratic that remains: $x^2 - 9x - 10 = (x - 10)(x + 1)$, and $x^3 - 9x^2 - 10x = x(x - 10)(x + 1)$. Set each factor equal to 0 and solve for $x$: $x = 0$; $x - 10 = 0$, $x = 10$; $x + 1 = 0$, $x = -1$. The roots of this equation are 0, −1, and 10.

## ▶ Undefined Expressions

As you saw in previous chapters, a fraction is undefined when its denominator is equal to 0. If the denominator of a fraction is a polynomial, factor it and set the factors equal to 0. The values that make the polynomial equal to 0 are the values that make the fraction undefined.

For what values of $x$ is the fraction $\frac{(x+49)}{(x^3 + 7x^2)}$ undefined?

Factor the denominator and set each factor equal to 0 to find the values of $x$ that make the fraction undefined. $x^3 + 7x^2 = x^2(x + 7)$, $x^2 = 0$, $x = 0$; $x + 7 = 0$, $x = -7$. The fraction is undefined when $x$ equals 0 or −7.

$$x^3 + 2x^2 + 5x^2 + 10x + 7x - 14$$
$$x^3 + 7x^2 + 3x - 14$$

## ▶ Practice

**1.** $-3x(x + 6)(x - 9) =$
   a. $-3x^3 + 6x - 54$
   b. $-x^3 + 3x^2 + 24x$
   c. $-3x^3 - 3x^2 - 54$
   d. $-3x^2 + 6x - 72$
   e. $-3x^3 + 9x^2 + 162x$

**2.** $(x^2 + 5x - 7)(x + 2) =$
   a. $x^3 - 3x^2 - 17x - 14$
   b. $x^3 + 5x^2 - 7x + 14$
   c. $x^3 + 7x^2 + 17x - 14$
   d. $x^3 + 7x^2 + 3x - 14$
   e. $2x^3 + 10x^2 - 14x$

$(x^2 - 9x + 18)(x - 1)$

$x^3 + x^2 - 9x^2 + 9x + 18x - 18$

**3.** $(x - 6)(x - 3)(x - 1) =$ $x^3 - 8x^2 + 27x - 18$

$x^3 - 8x^2 + 27x - 18$

   **a.** $x^3 - 18$

   **b.** $x^3 - 9x - 18$

   **c.** $x^3 - 8x^2 + 27x - 18$

   **d.** $x^3 - 10x^2 - 9x - 18$

   **e.** $x^3 - 10x^2 + 27x - 18$

**4.** What are the factors of $64x^3 - 16x$?

   **a.** $4(16x^3 - 16x)$

   **b.** $16(x^3 - x)$

   **c.** $16x(4x^2)$

   **d.** $16x(4x - 1)$

   **e.** $16x(2x - 1)(2x + 1)$

**5.** What are the factors of $2x^3 + 8x^2 - 192x$?

   **a.** $2(x - 8)(x + 12)$

   **b.** $2x(x - 8)(x + 12)$

   **c.** $x(2x - 8)(x + 24)$

   **d.** $2x(x + 16)(x - 12)$

   **e.** $2(x^2 - 8x)(x^2 + 12x)$

**6.** What is a root of $x(x - 1)(x + 1) = 27 - x$?

   **a.** $-9$

   **b.** $-1$

   **c.** $3$

   **d.** $1$

   **e.** $9$

**7.** The fraction $\frac{(x^2 + 8x)}{(x^3 - 64x)}$ is equivalent to

   **a.** $\frac{1}{(x - 8)}$

   **b.** $\frac{x}{(x - 8)}$

   **c.** $\frac{(x + 8)}{(x - 8)}$

   **d.** $x - 8$

   **e.** $x + 8$

**8.** The fraction $\frac{(x^2 + 6x + 5)}{(x^2 - 25x)}$ is equivalent to

   **a.** $\frac{(x + 1)}{x}$

   **b.** $\frac{(x + 1)}{(x + 5)}$

   **c.** $\frac{(x + 1)}{(x^2 - 5x)}$

   **d.** $\frac{(x + 1)}{(x^2 + 5)}$

   **e.** $\frac{(x + 1)}{(x^2 + 5x)}$

**9.** The fraction $\frac{(2x^2 + 4x)}{(4x^3 - 16x^2 - 48x)}$ is equivalent to

   **a.** $\frac{(x + 2)}{(x - 6)}$

   **b.** $\frac{x}{(x + 2)(x + 6)}$

   **c.** $\frac{1}{(2x - 12)}$

   **d.** $\frac{(x + 2)}{4x(x - 6)}$

   **e.** $\frac{2x(x + 2)}{(x - 6)}$

**10.** What is one value that makes the fraction $\frac{(x^2 - 25)}{(x^3 + 125)}$ undefined?

   **a.** $-25$

   **b.** $-5$

   **c.** $-1$

   **d.** $5$

   **e.** No values make the fraction undefined.

**11.** What values make the fraction $\frac{(x^2 + 7x + 12)}{(x^3 + 3x^2 - 4x)}$ undefined?

   **a.** $-4, 1$

   **b.** $-4, 0, 1$

   **c.** $-4, -1, 0$

   **d.** $-1, 0, 4$

   **e.** $0, 1, 4$

**12.** What is one value that makes the fraction $\frac{(x^2 + 11x + 30)}{(4x + 44x^2 + 120x)}$ undefined?

   **a.** −6

   **b.** −4

   **c.** −3

   **d.** −2

   **e.** −1

**13.** The cube of a number minus twice its square is equal to 80 times the number. If the number is greater than 0, what is the number?

   **a.** 4

   **b.** 5

   **c.** 8

   **d.** 10

   **e.** 20

**14.** Four times the cube of a number is equal to 48 times the number minus four times the square of the number. If the number is greater than 0, what is the number?

   **a.** 3

   **b.** 4

   **c.** 5

   **d.** 6

   **e.** 7

**15.** The product of 3 consecutive positive integers is equal to 56 more than the cube of the first integer. What is largest of these integers?

   **a.** 3

   **b.** 4

   **c.** 5

   **d.** 6

   **e.** 8

# Radicals and Exponents

## ▶ Radicals

In order to simplify some expressions and solve some equations, you will need to find the square or cube root of a number or variable. The **radical** symbol, $\sqrt{\ }$ , signifies the root of a value. The square root, or second root, of $x$ is equal to $\sqrt{x}$, or $\sqrt[2]{x}$. If there is no root number given, it is assumed that the radical symbol represents the square root of the number. The number under the radical symbol is called the **radicand**.

## ▶ Adding and Subtracting Radicals

Two radicals can be added or subtracted if they have the same radicand. To add two radicals with the same radicand, add the coefficients of the radicals and keep the radicand the same.

$$2\sqrt{2} + 3\sqrt{2} = 5\sqrt{2}$$

To subtract two radicals with the same radicand, subtract the coefficient of the second radical from the coefficient of the first radical and keep the radicand the same.

$$6\sqrt{5} - 4\sqrt{5} = 2\sqrt{5}$$

The expressions $\sqrt{3} + \sqrt{2}$ and $\sqrt{3} - \sqrt{2}$ cannot be simplified any further, since these radicals have different radicands.

## ▶ Multiplying Radicals

Two radicals can be multiplied whether or not they have the same radicand. To multiply two radicals, multiply the coefficients of the radicals and multiply the radicands.

$$(4\sqrt{6})(3\sqrt{7}) = 12\sqrt{42}, \text{ since } (4)(3) = 12 \text{ and } (\sqrt{6})(\sqrt{7}) = \sqrt{42}.$$

If two radicals of the same root with the same radicand are multiplied, the product is equal to the value of the radicand alone.

$(\sqrt{6})(\sqrt{6}) = 6$. Both radicals represent the same root, the square root, and both radicals have the same radicand, 6, so the product of $\sqrt{6}$ and $\sqrt{6}$ is 6.

## ▶ Dividing Radicals

Two radicals can be divided whether or not they have the same radicand. To divide two radicals, divide the coefficients of the radicals and divide the radicands.

$$\frac{(10\sqrt{15})}{(2\sqrt{3})} = 5\sqrt{5}, \text{ since } \frac{10}{2} = 5 \text{ and } \frac{\sqrt{15}}{\sqrt{3}} = \sqrt{5}.$$

Any radical divided by itself is equal to 1; $\frac{\sqrt{3}}{\sqrt{3}} = 1$.

## ▶ Simplifying a Single Radical

To simplify a radical such as $\sqrt{64}$, find the square root of 64. Look for a number that, when multiplied by itself, equals 64. Since $(8)(8) = 64$, the square root of 64 is 8; $\sqrt{64} = 8$; $\sqrt{64}$ is equal only to 8, not −8. The equation $x^2 = 64$ has two solutions, since both 8 and −8 square to 64, but the square root of a positive number must be a positive number.

However, most radicals cannot be simplified so easily. Many whole numbers and fractions do not have roots that are also whole numbers or fractions. You can simplify the original radical, but you will still have a radical in your answer.

To simplify a single radical, such as $\sqrt{32}$, find two factors of the radicand, one of which is a perfect square; $\sqrt{32} = (\sqrt{16})(\sqrt{2})$; $\sqrt{16}$ is a perfect square; the positive square root of 16 is 4. Therefore, $\sqrt{32} = (\sqrt{16})(\sqrt{2}) = 4\sqrt{2}$.

## ▶ Rationalizing Denominators of Fractions

An expression is not in simplest form if there is a radical in the denominator of a fraction. For example, the fraction $\frac{4}{\sqrt{3}}$ is not in simplest form. Multiply the top and bottom of the fraction by the radical in the denominator. Multiply $\frac{4}{\sqrt{3}}$ by $\frac{\sqrt{3}}{\sqrt{3}}$. Since $\frac{\sqrt{3}}{\sqrt{3}} = 1$, this will not change the value of the fraction. Since any radical multiplied by itself is equal to the radicand, $(\sqrt{3})(\sqrt{3}) = 3$; $(4)(\sqrt{3}) = 4\sqrt{3}$, so the fraction $\frac{4}{\sqrt{3}}$ in simplest form is $\frac{4\sqrt{3}}{3}$.

## ▶ Solving Equations with Radicals

Use the properties of adding, subtracting, multiplying, dividing, and simplifying radicals to help you solve equations with radicals. To remove a radical symbol from one side of an equation, you can raise both sides of the equation to a power. Remove a square root symbol from an equation by squaring both sides of the equation. Remove a cube root symbol from an equation by cubing both sides of the equation.

If $\sqrt{x} = 6$, what is the value of $x$?

To remove the radical symbol from the left side of the equation, square both sides of the equation. In other words, raise both sides of the equation to the power that is equal to the root of the radical. To remove a square root, or second root, raise both sides of the equation to the second power. To remove a cube root, or third root, raise both sides of the equation to the third power.

$\sqrt{x} = 6$, $(\sqrt{x})^2 = (6)^2$, $x = 36$; $\sqrt[3]{x} = 3$, $(\sqrt[3]{x})^3 = (3)^3$, $x = 27$

## ▶ Exponents

When a value, or base, is raised to a power, that power is the **exponent** of the base. The exponent of the term $4^2$ is two, and the base of the term is 4. The exponent is equal to the number of times a base is multiplied by itself; $4^2 = (4)(4)$; $2^6 = (2)(2)(2)(2)(2)(2)$.

Any value with an exponent of 0 is equal to 1; $1^0 = 1$, $10^0 = 1$, $x^0 = 1$.

Any value with an exponent of 1 is equal to itself; $1^1 = 1$, $10^1 = 10$, $x^1 = x$.

## ▶ Fractional Exponents

An exponent can also be a fraction. The numerator of the fraction is the power to which the base is being raised. The denominator of the fraction is the root of the base that must be taken. For example, the square root of a number can be represented as $x^{\frac{1}{2}}$, which means that $x$ must be raised to the first power ($x^1 = x$) and then the second, or square, root must be taken; $x^{\frac{1}{2}} = \sqrt{x}$.

$$4^{\frac{3}{2}} = (\sqrt{4})^3 = 2^3 = 8$$

It does not matter if you find the root (represented by the denominator) first, and then raise the result to the power (represented by the numerator), or if you find the power first and then take the root.

$$4^{\frac{3}{2}} = \sqrt{(4)^3} = \sqrt{64} = 8$$

## ▶ Negative Exponents

A base raised to a negative exponent is equal to the reciprocal of the base raised to the positive value of that exponent.

$$3^{-3} = \frac{1}{(3^3)}$$
$$x^{-2} = \frac{1}{(x^2)}$$

## ▶ Multiplying and Dividing Terms with Exponents

To multiply two terms with common bases, multiply the coefficients of the bases and add the exponents of the bases.

$$(3x^2)(7x^4) = 21x^6$$
$$(2x^{-5})(2x^3) = 4x^{-2}, \text{ or } \frac{4}{x^2}$$
$$(x^c)(x^d) = x^{c+d}$$

To divide two terms with common bases, divide the coefficients of the bases and subtract the exponents of the bases.

$$\frac{(27x^5)}{(9x)} = 3x^4$$

$$\frac{(2x^3)}{(8x^4)} = \frac{x^{-1}}{4}, \text{ or } \frac{1}{(4x)}$$

$$\frac{(x^c)}{(x^d)} = x^{c-d}$$

## ▶ Raising a Term with an Exponent to Another Exponent

When a term with an exponent is raised to another exponent, keep the base of the term and multiply the exponents.

$$(x^3)^3 = x^9$$
$$(x^c)^d = x^{cd}$$

If the term that is being raised to an exponent has a coefficient, be sure to raise the coefficient to the exponent as well.

$$(3x^2)^3 = 27x^6$$
$$(cx^3)^4 = c^4x^{12}$$

## ▶ Practice

**1.** $\sqrt{(32x^2)} =$
a. $4\sqrt{2x}$
b. $4x\sqrt{2}$
c. $4x\sqrt{8}$
d. $16x$
e. $16x\sqrt{2}$

*(handwritten: $32x^2$ / $16x$ $2x$ / $4x\sqrt{x}$)*

**2.** $a^3\sqrt{(a^3)} =$
a. $a^4\sqrt{a}$
b. $a^5$
c. $a^5\sqrt{a}$
d. $a^6$
e. $a^9$

**3.** $4\frac{\sqrt{g}}{\sqrt{4g}} =$
a. $2$
b. $4$
c. $\sqrt{g}$
d. $\frac{2\sqrt{g}}{g}$
e. $2\sqrt{g}$

**4.** $\frac{\sqrt[3]{(27y^3)}}{\sqrt{(27y^2)}} =$
a. $\frac{\sqrt{3}}{3}$
b. $\sqrt{3}$
c. $\frac{y\sqrt{3}}{3}$
d. $y$
e. $y\sqrt{3}$

**5.** $\frac{(\sqrt{(a^2b)})(\sqrt{(ab^2)})}{\sqrt{ab}} =$
a. $\frac{\sqrt{ab}}{ab}$
b. $\sqrt{ab}$
c. $ab$
d. $ab\sqrt{ab}$
e. $a^2b^2$

**6.** $\sqrt{\frac{m^3}{n^5}}^{-2} =$
a. $\frac{m^3}{n^5}$
b. $\frac{n^5}{m^3}$
c. $\frac{m^6}{n^{10}}$
d. $\frac{n^7}{m^5}$
e. $\frac{n^{10}}{m^6}$

**7.** $\left(\frac{(ab)^3}{b}\right)^4 =$

  **a.** $a^7$

  **b.** $a^{12}$

  **c.** $a^7b^6$

  **d.** $a^{12}b^8$

  **e.** $a^{12}b^{11}$

**8.** $((4g^2)^3(g^4))^{\frac{1}{2}} =$

  **a.** $8g^3$

  **b.** $8g^4$

  **c.** $8g^5$

  **d.** $8g^{10}$

  **e.** $8g^{12}$

**9.** $\dfrac{\sqrt{9pr}}{(pr)^{-\frac{3}{2}}} =$

  **a.** $\sqrt{3pr}$

  **b.** $\frac{3}{pr}$

  **c.** $3\sqrt{pr}$

  **d.** $3pr$

  **e.** $3p^2r^2$

**10.** $\dfrac{(\frac{x}{y})^2(\frac{y}{x})^{-2}}{xy} =$

  **a.** $\frac{1}{xy}$

  **b.** $\frac{x^3}{y^5}$

  **c.** $\frac{x^3}{y^3}$

  **d.** $x^3y^3$

  **e.** $x^5y^5$

**11.** If $a^{\frac{2}{3}} = 6$, then $a^{\frac{4}{3}} =$

  **a.** $\sqrt{3}$

  **b.** $\sqrt{6}$

  **c.** $3\sqrt{6}$

  **d.** $6\sqrt{6}$

  **e.** $36$

**12.** If $(\sqrt{p})^4 = q^{-2}$, and $q = -\frac{1}{3}$, what is one possible value of $p$?

  **a.** $-\frac{1}{3}$

  **b.** $\frac{1}{9}$

  **c.** $\frac{1}{3}$

  **d.** $3$

  **e.** $9$

**13.** What is the value of $(a\sqrt{b})^{-ab}$ if $a = \frac{1}{3}$ and $b = 9$?

  **a.** $\frac{1}{9}$

  **b.** $\frac{1}{3}$

  **c.** $1$

  **d.** $3$

  **e.** $9$

**14.** What is the value of $((xy)^y)^x$ if $x = 2$ and $y = -x$?

  **a.** $-4$

  **b.** $\frac{1}{256}$

  **c.** $\frac{1}{16}$

  **d.** $4$

  **e.** $16$

**15.** If $g\sqrt{108} = \frac{\sqrt{3}}{g}$, what is a value of $g$?

    **a.** $\frac{1}{36}$

    **b.** $\frac{1}{6}$

    **c.** $\frac{\sqrt{6}}{6}$

    **d.** $\sqrt{6}$

    **e.** 6

**16.** If $(c\sqrt{d})^2 = 48$ and $c = 2$, what is the value of $d$?

    **a.** $2\sqrt{3}$

    **b.** $2\sqrt{6}$

    **c.** 6

    **d.** $4\sqrt{6}$

    **e.** 12

**17.** If $\left(\frac{\sqrt{n}}{n^{-\frac{1}{2}}}\right)m = 5$ what is the value of $m$ in terms of $n$?

    **a.** $\frac{n}{5}$

    **b.** $\frac{5}{n^{\frac{1}{4}}}$

    **c.** $5n^{\frac{1}{4}}$

    **d.** $\frac{5}{n}$

    **e.** $5\sqrt{n}$

**18.** What is the value of $(x^{-y})(2x^y)(3y^x)$ if $x = 2$ and $y = -2$?

    **a.** 6

    **b.** 8

    **c.** 12

    **d.** 24

    **e.** 384

**19.** If $n = 20$, what is the value of $\frac{\sqrt{n+5}}{\sqrt{n}}\left(\frac{n}{2}\sqrt{5}\right)$?

    **a.** 5

    **b.** $\frac{5\sqrt{5}}{2}$

    **c.** 10

    **d.** $5\sqrt{5}$

    **e.** 25

**20.** If $a$ is positive, and $a^2 = b = 4$, what is the value of $\left(\frac{b\sqrt{b}}{a^4}\right)^a$?

    **a.** $\frac{1}{1,024}$

    **b.** $\frac{1}{32}$

    **c.** $\frac{1}{16}$

    **d.** $\frac{1}{4}$

    **e.** $\frac{1}{2}$

# Sequences

A **sequence** is a series of terms in which each term in the series is generated using a rule. Each value in the sequence is a called a **term**. The rule of a sequence could be "each term is twice the previous term" or "each term is 4 more than the previous term." On the SAT, the rule of the sequence will always be given to you. The first term of the sequence is referred to as the first term (not the zeroth term).

## ▶ Arithmetic Sequences

An **arithmetic sequence** is a series of terms in which the difference between any two consecutive terms in the sequence is always the same. For example, 2, 6, 10, 14, 18, . . . is an arithmetic sequence. The difference between any two consecutive terms is always 4.

Each term in the sequence below is five less than the previous term. What is the next term in the sequence?

38, 33, 28, 23, . . .

To find the next term in the sequence, take the last term given and subtract 5, since the rule of the sequence is "each term in the sequence is five less than the previous term"; $23 - 5 = 18$, so the next term in the sequence is 18.

## ▶ Geometric Sequences

A **geometric sequence** is a series of terms in which the ratio between any two consecutive terms in the sequence is always the same. For example, $1, 3, 9, 27, 81, \ldots$ is a geometric sequence. The ratio between any two consecutive terms is always 3—each term is three times the previous term.

Each term in the sequence below is six times the previous term. What is the value of $x$?

$2, 12, x, 432, \ldots$

To find the value of $x$, the third term in the sequence, multiply the second term in the sequence, 12, by 6, since every term is six times the previous term; $(12)(6) = 72$. The third term in the sequence is 72. You can check your answer by multiplying 72 by 6: $(72)(6) = 432$, the fourth term in the sequence.

Sometimes you will be asked for the 20th, 50th, or 100th term of a sequence. It would be unreasonable in many cases to evaluate that term, but you can represent that term with an expression.

Each term in the sequence below is four times the previous term. What is the 100th term of the sequence?

$3, 12, 48, 192, \ldots$

Write each term of the sequence as a product; 3 is equal to $4^0 \times 3$; 12 is equal to $4^1 \times 3$, 48 is equal to $4^2 \times 3$, and 192 is equal to $4^3 \times 3$. Each term in the sequence is equal to 4 raised to an exponent, multiplied by 3. For each term, the value of the exponent is one less than the position of the term in the sequence. The fourth term in the sequence, 192, is equal to 4 raised to one less than four (3), multiplied by 3. Therefore, the 100th term of the sequence is equal to 4 raised to one less than 100 (99), multiplied by 3. The 100th term is equal to $4^{99} \times 3$.

## ▶ Combination Sequences

Some sequences are defined by rules that are a combination of operations. The terms in these sequences do not differ by a constant value or ratio. For example, each number in a sequence could be generated by the rule "double the previous term and add one": $5, 11, 23, 47, 95, \ldots$

Each term in the sequence below is one less than four times the previous term. What is the next term in the sequence?

$1, 3, 11, 43, \ldots$

Take the last given term in the sequence, 43, and apply the rule; $4(43) - 1 = 172 - 1 = 171$.

# ▶ Practice

**1.** Each term in the sequence below is nine less than the previous term. What is the ninth term of the sequence?

$$101, 92, 83, 74, \ldots$$

   **a.** 9
   **b.** 20
   **c.** 29
   **d.** 38
   **e.** 119

**2.** Each term in the sequence below is $\frac{3}{2}$ more than the previous term. What is the eighth term of the sequence?

$$6, 7\frac{1}{2}, 9, 10\frac{1}{2}, \ldots$$

   **a.** 15
   **b.** $15\frac{1}{2}$
   **c.** 16
   **d.** $16\frac{1}{2}$
   **e.** 18

**3.** Each term in the sequence below is seven less than the previous term. What is the value of $x - y$?

$$12, 5, x, y, -16, \ldots$$

   **a.** $-11$
   **b.** $-9$
   **c.** $-2$
   **d.** 5
   **e.** 7

**4.** Each term in the sequence below is six more than the previous term. What is the value of $x + z$?

$$x, y, z, 7, 13, \ldots$$

   **a.** $-16$
   **b.** $-12$
   **c.** $-10$
   **d.** $-6$
   **e.** $-4$

**5.** Each term in the sequence below is $\frac{1}{3}$ more than the previous term. What is the value of $a + b + c + d$?

$$2, a, b, 3, c, d, 4, \ldots$$

   **a.** 10
   **b.** 11
   **c.** $11\frac{1}{3}$
   **d.** $11\frac{2}{3}$
   **e.** 12

**6.** Each term in the sequence below is $-2$ times the previous term. What is the seventh term of the sequence?

$$3, -6, 12, -24, \ldots$$

   **a.** $-96$
   **b.** 192
   **c.** 384
   **d.** 768
   **e.** 1,536

**7.** Each term in the sequence below is $\frac{2}{3}$ times the previous term. What is the seventh term of the sequence?

$$18, 12, 8, \frac{16}{3}, \ldots$$

   **a.** $\frac{128}{81}$
   **b.** $\frac{64}{27}$
   **c.** $\frac{128}{27}$
   **d.** $\frac{64}{9}$
   **e.** $\frac{128}{3}$

**8.** Each term in the sequence below is five times the previous term. What is the 20th term of the sequence?

$$\frac{1}{125}, \frac{1}{25}, \frac{1}{5}, 1, \ldots$$

   **a.** $5^{16}$
   **b.** $5^{17}$
   **c.** $5^{19}$
   **d.** $5^{20}$
   **e.** $5^{21}$

**9.** Each term in the sequence below is –4 times the previous term. What is the value of $xy$?

$x, y, -64, 256, \ldots$

a. –256

b. –64

c. –16

d. 16

e. 64

**10.** Each term in the sequence below is three times the previous term. What is the product of the 100th and 101st terms of the sequence?

$1, 3, 9, 27, \ldots$

a. $3^{199}$

b. $3^{200}$

c. $3^{201}$

d. $3^{300}$

e. $3^{9,900}$

**11.** Each term in the sequence below is two less than three times the previous term. What is the next term of the sequence?

$-1, -5, -17, -53, \ldots$

a. –162

b. –161

c. –159

d. –158

e. –157

**12.** Each term in the sequence below is nine more than $\frac{1}{3}$ the previous term. What is the value of $y - x$?

$81, 36, x, y, \ldots$

a. –8

b. –7

c. –5

d. 7

e. 8

**13.** Each term in the sequence below is 20 less than five times the previous term. What is the value of $x + y$?

$x, 0, y, -120, \ldots$

a. –40

b. –28

c. –24

d. –20

e. –16

**14.** Each term in the sequence below is two less than $\frac{1}{2}$ the previous term. What term of the sequence will be the first term to be a negative number?

$256, 126, 61, 28.5, \ldots$

a. seventh

b. eighth

c. ninth

d. tenth

e. eleventh

**15.** Each term in the sequence below is 16 more than –4 times the previous term. What is the value of $x + y$?

$x; y; -80; 336; -1,328; \ldots$

a. –32

b. –16

c. 20

d. 22

e. 26

**16.** Each term in the sequence below is equal to the sum of the two previous terms.

$\ldots a, b, c, d, e, f, \ldots$

All of the following are equal to the value of $d$ EXCEPT

a. $e - c$

b. $b + c$

c. $a + 2b$

d. $e - 2b$

e. $f - e$

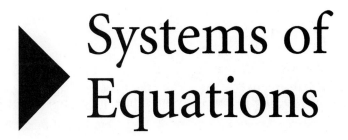

# Systems of Equations

**A** system of equations is a group of two or more equations in which the variables have the same values in each equation. If the number of equations is greater than or equal to the number of variables, you can find the value of each variable. The SAT typically features a system of two equations with two variables. There are two techniques that can be used to solve systems of equations: substitution and combination (sometimes referred to as elimination).

## ▶ Substitution

To solve a system of equations using substitution, take one equation and rewrite it so that you have the value of one variable in terms of the other. Then, substitute that expression into the second equation and solve.

Solve this system of equations for $x$ and $y$:

$2x + 2y = 4$
$3y - 4x = 13$

Take the first equation and rewrite it so that you have the value of $x$ in terms of $y$ (you could also choose to rewrite the equation with the value of $y$ in terms of $x$). Subtract $2y$ from both sides of the equation and divide by 2:

$$2x + 2y = 4$$
$$2x + 2y - 2y = 4 - 2y$$
$$2x = 4 - 2y$$
$$\frac{2x}{2} = \frac{(4 - 2y)}{2}$$
$$x = 2 - y$$

Now that you have the value of $x$ in terms of $y$, substitute this expression for $x$ in the second equation and solve for $y$:

$$3y - 4(2 - y) = 13$$
$$3y - 8 + 4y = 13$$
$$7y - 8 = 13$$
$$7y = 21$$
$$y = 3$$

Now that you have the value of $y$, substitute 3 for $y$ in either equation and solve for $x$:

$$x = 2 - (3)$$
$$x = -1$$

The solution to this system of equations is $(-1,3)$. Alternatively, you could have used the second equation to rewrite $x$ in terms of $y$ (or $y$ in terms of $x$), but that would have been more cumbersome, since the expressions would have involved fractions. There are many ways to approach solving a system of equations, and some methods work more easily that others, depending on the equations in the system.

## ▶ Combination

To solve a system of equations using combination, multiply one equation by a constant, and then add it to the other equation to eliminate a variable. The same system of equations you just saw could also be solved using combination.

$$2x + 2y = 4$$
$$3y - 4x = 13$$

Adding one equation to the other won't eliminate either variable. But, if the first equation is multiplied by 2 and then added to the second equation, the $x$ terms will drop out:

$(2)(2x + 2y = 4) = 4x + 4y = 8$
$(4x + 4y = 8)$
$\underline{+ (3y - 4x = 13)}$
$\qquad 7y = 21$

Divide both sides of the equation by 7, and you can see that $y = 3$, the same answer found using substitution. Now, substitute the value of $y$ into either equation to find the value of $x$:

$3(3) - 4x = 13$
$9 - 4x = 13$
$-4x = 4$
$x = -1$

After finding the value of $y$ using substitution, you found the value of $x$ using the equation $2x + 2y = 4$. After find the value of $y$ using combination, you found the value of $x$ using the equation $3y - 4x = 13$. In either example, you could have used either equation, and you still would have arrived at the same answer.

When faced with a system of equations, you can use either substitution or combination to find the solution. Both methods will work, although one may involve less work than the other.

## ▶ Practice

**1.** Given the equations below, what is the value of $x$?

$2x + y = 6$
$\frac{y}{2} + 4x = 12$

   **a.** −2
   **b.** 0
   **c.** 1
   **d.** 3
   **e.** 6

**2.** Given the equations below, what is the value of $b$?

$5a + 3b = -2$
$5a - 3b = -38$

   **a.** −6
   **b.** −4
   **c.** 6
   **d.** 12
   **e.** 13

**3.** Given the equations below, what is one possible value of $y$?

$xy = 32$
$2x - y = 0$

   **a.** −8
   **b.** −2
   **c.** 2
   **d.** 4
   **e.** 16

**4.** Given the equations below, what is the value of $x$?

$3(x + 4) - 2y = 5$
$2y - 4x = 8$

   **a.** −2
   **b.** −1
   **c.** 1
   **d.** 13
   **e.** 15

**5.** Given the equations below, what is the value of $b$?

$$-7a + \frac{b}{4} = 25$$
$$b + a = 13$$

*handwritten:* $-28a + b = 100$

$b = 100 + 28$

$100 + 28a = 13 - a$

$100 + 27a = 13$

$27a = 87$

   **a.** −3
   **b.** 4
   **c.** 12
   **d.** 13
   **e.** 16

**6.** Given the equations below, what is the value of $y$?

$$3x + 7y = 19$$
$$\frac{4y}{x} = 1$$

   **a.** $\frac{1}{4}$
   **b.** $\frac{1}{2}$
   **c.** 1
   **d.** 2
   **e.** 4

**7.** Given the equations below, what is the value of $n$?

$$2(m + n) + m = 9$$
$$3m - 3n = 24$$

   **a.** −5
   **b.** −3
   **c.** 3
   **d.** 5
   **e.** 8

**8.** Given the equations below, what is the value of $b$?

$$9a - 2(b + 4) = 30$$
$$4\frac{1}{2}a - 3b = 3$$

   **a.** 2
   **b.** 4
   **c.** 6
   **d.** 8
   **e.** 10

**9.** Given the equations below, what is one possible value of $p$?

$$4pq - 6 = 10$$
$$4p - 2q = -14$$

   **a.** −2
   **b.** $\frac{1}{2}$
   **c.** 1
   **d.** 2
   **e.** 8

**10.** Given the equations below, what is the value of $a$?

$$7(2a + 3b) = 56$$
$$b + 2a = -4$$

   **a.** −5
   **b.** −4
   **c.** −2
   **d.** 4
   **e.** 6

**11.** Given the equations below, what is the value of $y$?

$$\frac{1}{2}x + 6y = 7$$
$$-4x - 15y = 10$$

   **a.** −10
   **b.** $-\frac{1}{2}$
   **c.** 2
   **d.** 5
   **e.** 6

**12.** Given the equations below, what is one possible value of $m$?

$$m(n + 1) = 2$$
$$m - n = 0$$

   **a.** −2
   **b.** −1
   **c.** 0
   **d.** $\frac{1}{2}$
   **e.** 2

**13.** Given the equations below, what is the value of $\frac{c}{d}$?

$$\frac{c-d}{5} - 2 = 0$$
$$c - 6d = 0$$

**a.** 2
**b.** 6
**c.** 8
**d.** 12
**e.** 14

**14.** Given the equations below, what is the value of $a + b$?

$$4a + 6b = 24$$
$$6a - 12b = -6$$

**a.** 2
**b.** 3
**c.** 4
**d.** 5
**e.** 6

**15.** Given the equations below, what is the value of $x - y$?

$$\frac{x}{3} - 2y = 14$$
$$2x + 6y = -6$$

**a.** −7
**b.** 5
**c.** 7
**d.** 12
**e.** 17

**16.** Given the equations below, what is the value of $xy$?

$$-5x + 2y = -51$$
$$-x - y = -6$$

**a.** −27
**b.** −18
**c.** −12
**d.** −6
**e.** −3

**17.** Given the equations below, what is the value of $\frac{n}{m}$?

$$m - 6(n + 2) = -8$$
$$6n + m = 16$$

**a.** −10
**b.** −9
**c.** $\frac{1}{10}$
**d.** 9
**e.** 10

**18.** Given the equations below, what is the value of $y - x$?

$$3x + 4 = -5y + 8$$
$$9x + 11y = -8$$

**a.** −12
**b.** −2
**c.** 2
**d.** 12
**e.** 14

**19.** Given the equations below, what is the value of $a + b$?

$$\frac{1}{2}(a + 3) - b = -6$$
$$3a - 2b = -5$$

**a.** 5
**b.** 15
**c.** 20
**d.** 25
**e.** 45

**20.** Given the equations below, what is the value of $ab$?

$$10b - 9a = 6$$
$$b - a = 1$$

**a.** −12
**b.** −7
**c.** 1
**d.** 7
**e.** 12

**21.** Given the equations below, what is the value of $x - y$?

$$\frac{x+y}{3} = 8$$
$$2x - y = 9$$

a. $-24$

b. $-2$

c. $0$

d. $1$

e. $2$

**22.** Given the equations below, what is the value of $\frac{x}{y}$?

$$4x + 6 = -3y$$
$$-2x + 3 = y + 9$$

a. $-6$

b. $-1$

c. $0$

d. $1$

e. $6$

**23.** Given the equations below, what is the value of $(p + q)^2$?

$$8q + 15p = 26$$
$$-5p + 2q = 24$$

a. $4$

b. $5$

c. $25$

d. $49$

e. $81$

**24.** Given the equations below, what is the value of $(y - x)^2$?

$$9(x - 1) = 2 - 4y$$
$$2y + 7x = 3$$

a. $1$

b. $4$

c. $16$

d. $25$

e. $36$

**25.** Given the equations below, what is the value of $\sqrt{\frac{a}{b}}$?

$$\frac{a}{2} = b + 1$$
$$3(a - b) = -21$$

a. $\frac{4}{9}$

b. $\frac{2}{3}$

c. $\frac{3}{4}$

d. $\frac{4}{3}$

e. $\frac{3}{2}$

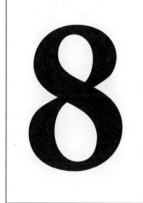

# Functions, Domain, and Range

## ▶ Functions

A **function** is an equation with one input (variable) in which each unique input value yields no more than one output. The set of elements that make up the possible inputs of a function is the **domain** of the function. The set of elements that make up the possible outputs of a function is the **range** of the function.

A function commonly takes the form $f(x) = x + c$, where $x$ is a variable and $c$ is a constant. The values for $x$ are the domain of this function. The values of $f(x)$ are the range of the function.

If $f(x) = 5x + 2$, what is $f(3)$?

To find the value of a function given an input, substitute the given input for the variable. $f(3) = 5(3) + 2 = 15 + 2 = 17$.

## ▶ Vertical Line Test

An equation is a function if it passes the **vertical line test**. To test if an equation is a function, draw the graph of the equation. If you can draw a vertical line through any point of the graph and the line crosses the graph no more than once, the equation is a function. If the vertical line crosses the graph more than once for any point, the equation is not a function.

Is $y = x^2$ a function?

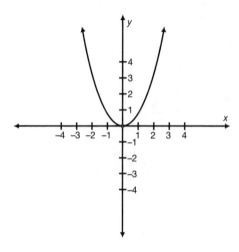

For every value of $x$, only one $y$ value is returned. There is no value of $x$ that can be substituted into the equation that can yield more than one different value for $y$.

Look at the graph of $y = x^2$ at left.

If a vertical line is drawn through any point on the graph, it will cross the graph of $y = x^2$ only once.

The equation $y = x^2$ passes the vertical line test, so it is a function.

Is $x = y^2$ a function?

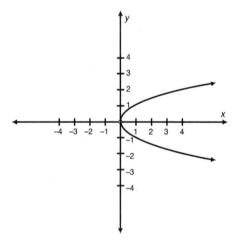

For every value of $x$ that is greater than 0, two $y$ values are returned. If a vertical line is drawn through any point to the right of the $y$-axis, it will cross the graph of the equation $x = y^2$ in two places.

This equation fails the vertical line test, so it is not a function.

Always be wary of equations that are written in the form $x =$ rather than $y =$. For example, the equation $x = c$, where $c$ is any constant, is not a function for any value of $c$, since that is the graph of a vertical line. A vertical line is not a function.

## ▶ Horizontal Line Test

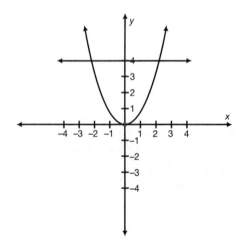

The **horizontal line test** can be used to determine how many different $x$ values, or inputs, return the same $f(x)$ value. Remember, a function cannot have one input return two or more outputs, but it can have more than one input return the same output. For example, the function $f(x) = x^2$ is a function, because no $x$ value can return two or more $f(x)$ values, but more than one $x$ value can return the same $f(x)$ value. Both $x = 2$ and $x = -2$ make $f(x) = 4$. To find how many values make $f(x) = 4$, draw a horizontal line through the graph of the function where $f(x)$, or $y$, = 4.

You can see that the line $y = 4$ crosses the graph of $f(x) = x^2$ in exactly two places. Therefore, the horizontal line test proves that there are two values for $x$ that make $f(x) = 4$.

## ▶ Domain

The function $f(x) = 3x$ has a domain of all real numbers. Any real number can be substituted for $x$ in the equation and the value of the function will be a real number.

The function $f(x) = \frac{2}{x-4}$ has a domain of all real numbers excluding 4. If $x = 4$, the value of the function would be $\frac{2}{0}$, which is undefined. In earlier chapters, you found the values for an expression that make the expression undefined. In a function, the values that make a part of the function undefined are the values that are NOT in the domain of the function.

What is the domain of the function $f(x) = \sqrt{x}$?

The square root of a negative number is an imaginary number, so the value of $x$ must not be less than 0. Therefore, the domain of the function is $x \geq 0$.

## ▶ Range

As you just saw, the function $f(x) = 3x$ has a domain of all real numbers. If any real number can be substituted for $x$, $3x$ can yield any real number. The range of this function is also all real numbers.

Although the domain of the function $f(x) = \frac{2}{x-4}$ is all real numbers excluding 4, the range of the function is all real numbers excluding 0, since no value for $x$ can make $f(x) = 0$.

What is the range of the function $f(x) = \sqrt{x}$?

You already found the domain of the function to be $x \geq 0$. For all values of $x$ greater than or equal to 0, the function will return values greater than or equal to 0.

## ▶ Nested Functions

Given the definitions of two functions, you can find the result of one function (given a value) and place it directly into another function. For example, if $f(x) = 5x + 2$ and $g(x) = -2x$, what is $f(g(x))$ when $x = 3$?

Begin with the innermost function: Find $g(x)$ when $x = 3$. In other words, find $g(3)$. Then, substitute the result of that function for $x$ in $f(x)$; $g(3) = -2(3) = -6$, $f(-6) = 5(-6) + 2 = -30 + 2 = -28$. Therefore, $f(g(x)) = -28$ when $x = 3$.

What is the value of $g(f(x))$ when $x = 3$?

Start with the innermost function—this time, it is $f(x)$; $f(3) = 5(3) + 2 = 15 + 2 = 17$. Now, substitute 17 for $x$ in $g(x)$; $g(17) = -2(17) = -34$. When $x = 3$, $f(g(x)) = -28$ and $g(f(x)) = -34$.

## ▶ Newly Defined Symbols

A symbol can be used to represent one or more operations. For a particular question on the SAT, a symbol such as # may be given a certain definition, such as "$m\#n$ is equivalent to $m^2 + n$." You may be asked to find the value of the function given the values of $m$ and $n$, or you may be asked to find an expression that represents the function.

If $m\#n$ is equivalent to $m^2 + n$, what is the value of $m\#n$ when $m = 2$ and $n = -2$?

Substitute the values of $m$ and $n$ into the definition of the symbol. The definition of the function states that the term before the # symbol should be squared and added to the term after the # symbol. When $m = 2$ and $n = -2$, $m^2 + n = (2)^2 + (-2) = 4 - 2 = 2$.

If $m\#n$ is equivalent to $m^2 + 2n$, what is the value of $n\#m$?

The definition of the function states that the term before the # symbol should be squared and added to twice the term after the # symbol. Therefore, the value of $n\#m = n^2 + 2m$. Watch your variables carefully. The definition of the function is given for $m\#n$, but the question asks for the value of $n\#m$.

If $m\#n$ is equivalent to $m + 3n$, what is the value of $n\#(m\#n)$?

Begin with the innermost function, $m\#n$. The definition of the function states that the term before the # symbol should be added to three times the term after the # symbol. Therefore, the value of $m\#n = m + 3n$. That expression, $m + 3n$, is now the term after the # symbol: $n\#(m + 3n)$. Look again at the definition of the function. Add the term before the # symbol to three times the term after the # symbol. Add $n$ to three times $(m + 3n)$: $n + 3(m + 3n) = n + 3m + 9n = 3m + 10n$.

## ▶ Practice

**1.** The graph of $f(x)$ is shown below. For how many values is $f(x) = 2$?

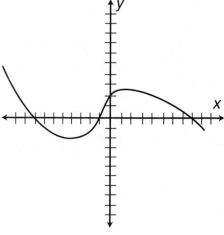

   **a.** 1
   **b.** 2
   **c.** 3
   **d.** 4
   **e.** 5

**2.** The graph of $f(x)$ is shown below. For how many values is $f(x) = 3$?

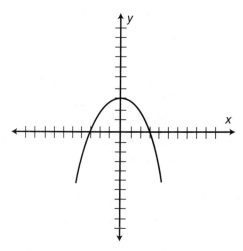

   **a.** 0
   **b.** 1
   **c.** 2
   **d.** 3
   **e.** 4

**3.** The graph of $f(x)$ is shown below. For how many values is $f(x) = 0$?

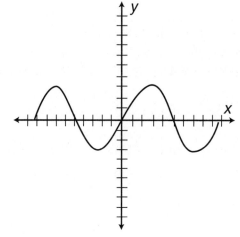

   **a.** 2
   **b.** 3
   **c.** 4
   **d.** 5
   **e.** 8

**4.** The graph of $f(x)$ is shown below. For how many values is $f(x) = -5$?

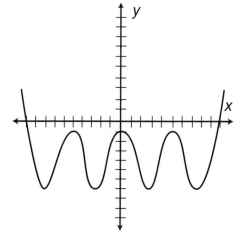

   **a.** 0
   **b.** 2
   **c.** 4
   **d.** 5
   **e.** 8

**5.** The graph of $f(x)$ is shown below. For how many values is $f(x) = -2$?

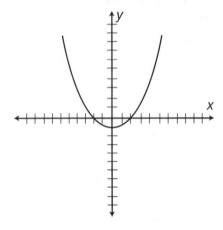

　　a. 0
　　b. 1
　　c. 2
　　d. 3
　　e. 4

**6.** If $f(x) = 2x - 1$ and $g(x) = x^2$, what is the value of $f(g(-3))$?
　　a. −7
　　b. 2
　　c. 9
　　d. 17
　　e. 49

**7.** If $f(x) = 3x + 2$ and $g(x) = 2x - 3$, what is the value of $g(f(-2))$?
　　a. −19
　　b. −11
　　c. −7
　　d. −4
　　e. −3

**8.** If $f(x) = 2x + 1$ and $g(x) = x - 2$, what is the value of $f(g(f(3)))$?
　　a. 1
　　b. 3
　　c. 5
　　d. 7
　　e. 11

**9.** If $f(x) = 6x + 4$ and $g(x) = x^2 - 1$, what is the value of $g(f(x))$?
　　a. $6x^2$　2
　　b. $36x^2 + 16$
　　c. $36x^2 + 48x + 15$
　　d. $36x^2 + 48x + 16$
　　e. $6x^3 + 4x^2 - 6x - 4$

**10.** If $f(x) = 4 - 2x^2$ and $g(x) = 2\sqrt{x}$, what is the value of $f(g(x))$?
　　a. $4 - 16x$
　　b. $4 - 8x$
　　c. $4 - 4x$
　　d. $4 - 2x$
　　e. $4 - x$

**11.** If $w@z$ is equivalent to $3w - z$, what is the value of $(w@z)@z$?
　　a. $3w - 2z$
　　b. $6w - 2z$
　　c. $9w - 4z$
　　d. $9w - 3z$
　　e. $9w - 2z$

**12.** If $p\&q$ is equivalent to $\frac{p}{q} + pq$, what is the value of $q\&p$ when $p = 4$ and $q = -2$?
　　a. −10
　　b. −8.5
　　c. −7.5
　　d. 4
　　e. 16

**13.** If $j\%k$ is equivalent to $k^j$, what is the value of $k\%(j\%k)$?
　　a. $k^{jk}$
　　b. $j^{kj}$
　　c. $k^{(j+k)}$
　　d. $j^{(k+j)}$
　　e. $k^j j^k$

**14.** If $a?b$ is equivalent to $\frac{b-a}{a+b}$, what is the value of $a?(a?b)$ when $a = 6$ and $b = -5$?

   **a.** $-11$
   **b.** $-\frac{7}{5}$
   **c.** $-1$
   **d.** $1$
   **e.** $\frac{17}{5}$

**15.** If $x{\wedge}y$ is equivalent to $y^2 - x$, what is the value of $y{\wedge}(y{\wedge}y)$?

   **a.** $y^2 - 2x$
   **b.** $y^2 - 2y$
   **c.** $y^4 - 2xy^2 + x^2 - y$
   **d.** $y^4 - 2y^3 + y^2 - y$
   **e.** $y^4 - y^3 + y^2 - y$

**16.** What is the domain of the function $f(x) = \frac{1}{(x^2 - 9)}$?

   **a.** all real numbers excluding 0
   **b.** all real numbers excluding 3 and –3
   **c.** all real numbers greater than 9
   **d.** all real numbers greater than or equal to 9
   **e.** all real numbers greater than 3 and less than –3

**17.** What is the range of the function $f(x) = x^2 - 4$?

   **a.** all real numbers excluding 0
   **b.** all real numbers excluding 2 and –2
   **c.** all real numbers greater than or equal to 0
   **d.** all real numbers greater than or equal to 4
   **e.** all real numbers greater than or equal to –4

**18.** Which of the following is true of $f(x) = \frac{-x^2}{2}$?

   **a.** The range of the function is all real numbers less than or equal to 0.
   **b.** The range of the function is all real numbers less than 0.
   **c.** The range of the function is all real numbers greater than or equal to 0.
   **d.** The domain of the function is all real numbers greater than or equal to 0.
   **e.** The domain of the function is all real numbers less than or equal to $\sqrt{2}$.

**19.** Which of the following is true of $f(x) = \sqrt{4x - 1}$?

   **a.** The domain of the function is all real numbers greater than $\frac{1}{4}$ and the range is all real numbers greater than 0.
   **b.** The domain of the function is all real numbers greater than or equal to $\frac{1}{4}$ and the range is all real numbers greater than 0.
   **c.** The domain of the function is all real numbers greater than or equal to $\frac{1}{4}$ and the range is all real numbers greater than or equal to 0.
   **d.** The domain of the function is all real numbers greater than 0 and the range is all real numbers greater than or equal to $\frac{1}{4}$.
   **e.** The domain of the function is all real numbers greater than or equal to 0 and the range is all real numbers greater than or equal to $\frac{1}{4}$.

**20.** Which of the following is true of $f(x) = \frac{-1}{\sqrt{x-5}}$?

   **a.** The domain of the function is all real numbers excluding 5 and the range is all real numbers less than or equal to 0.
   **b.** The domain of the function is all real numbers greater than 5 and the range is all real numbers less than or equal to 0.
   **c.** The domain of the function is all real numbers greater than 5 and the range is all real numbers less than 0.
   **d.** The domain of the function is all real numbers greater than or equal to 5 and the range is all real numbers less than 0.
   **e.** The domain of the function is all real numbers greater than or equal to 5 and the range is all real numbers less than or equal to 0.

Use the diagrams below to answer questions 21–25.

A

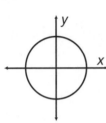

$x^2 + y^2 = 4$

B

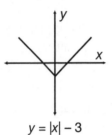

$y = |x| - 3$

C

$y = \sqrt{x}$

D

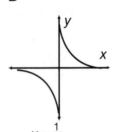

$y = \frac{1}{x}$

E

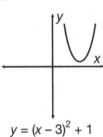

$y = (x - 3)^2 + 1$

**21.** Which of the coordinate planes shows the graph of an equation that is not a function?
  a. A
  b. B
  c. C
  d. D
  e. A, D

**22.** Which of the coordinate planes shows the graph of a function that has a range that contains negative values?
  a. A
  b. B
  c. D
  d. B, D
  e. A, B, D

**23.** Which of the coordinate planes shows the graph of a function that has a domain of all real numbers?
  a. B
  b. D
  c. E
  d. B, D
  e. B, $\sqrt{E}$

**24.** Which of the coordinate planes shows the graph of a function that has the same range as its domain?
  a. B, C
  b. C, D
  c. B, D
  d. B, E
  e. D, E

**25.** Of the equations graphed on the coordinate planes, which function has the smallest range?
  a. A
  b. B
  c. C
  d. D
  e. E

# Angles

## ▶ Naming an Angle

A **ray** is a line with one endpoint. Two rays that meet at their endpoints form an **angle**. The **vertex** of the angle is the place where the two endpoints meet.

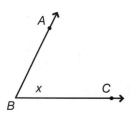

The vertex of an angle determines its name. An angle is named by either its vertex alone, its label (if it has one), or by the rays and the vertex that form the angle. The angle at left could be named *B*, because that is its vertex, *x*, because that is its label, or *ABC*, because it is formed by rays *AB* and *BC*. The angle could also be named *CBA*, since it is formed by rays *BC* and *BA*. The order in which rays or line segments are named does not matter, but the vertex of the angle must always be the middle letter.

## ► Types of Angles

**Acute angles** are angles that measure less than 90°. **Obtuse angles** are angles that measure greater than 90°. **Right angles** are angles that measure exactly 90°. **Straight angles** measure exactly 180°. It is important to remember that there are 180° in a straight line.

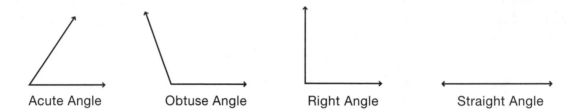

| Acute Angle | Obtuse Angle | Right Angle | Straight Angle |

## ► Complementary, Supplementary, and Adjacent Angles

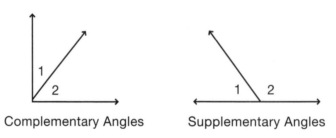

Complementary Angles          Supplementary Angles

The measures of **complementary angles** add to 90° and the measures of **supplementary angles** add to 180°. Two angles that combine to form a right angle.

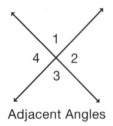

Adjacent Angles

**Adjacent angles** are two angles that share a common vertex and a common ray. The complementary angles shown are also adjacent angles, as are the two supplementary angles.

## ► Vertical Angles and Alternating Angles

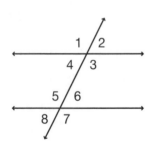

Intersecting lines form **vertical angles**. When two lines intersect, the angles formed on either side of the vertex (the point of intersection) are called opposite, or vertical, angles. Vertical angles are always **congruent**, or equal in measure. Angles 1 and 3 are congruent, and angles 2 and 4 are congruent.

Parallel lines cut by a transversal (a third line that is not parallel to the first two lines) also form vertical angles. Sets of vertical angles are called **alternating angles**. In the diagram at left, the parallel lines and the transversal form eight angles. The angles labeled 1 and 3 are vertical angles, as are the angles labeled 5 and 7. All four of these angles are congruent. Notice how these congruent angles alternate back and forth across the transversal as you look down the transversal. These vertical angles are alternating angles. Angles 1, 3, 5, and 7 are all alternating, and therefore, congruent. In the same way, angles 2, 4, 6, and 8 are all alternating, and they are all congruent.

Angles 1 and 2 are not alternating, but they do form a line. These two angles are supplementary. Every odd-numbered angle is congruent to every other odd-numbered angle, and every odd-numbered angle is supplementary, and adjacent, to every even-numbered angle. In the same way, every even-numbered angle is congruent to every other even-numbered angle, and every even-numbered angle is supplementary, and adjacent, to every odd-numbered angle.

## ▶ Practice

Use the diagram below to answer questions 1–5. The diagram is not to scale.

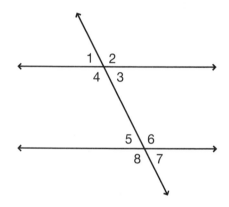

**1.** If the measure of angle 2 is equal to $12x + 10$ and the measure of angle 8 is equal to $7x - 1$, what is the measure of angle 2?
a. 9°
b. 62°
c. 108°
d. 118°
e. cannot be determined

*handwritten:* $12x + 10 = 7x - 1$
$5x + 10 = -1$
$5x = 9$

**2.** If the measure of angle 5 is five times the measure of angle 6, what is the measure of angle 5?
a. 30°
b. 36°
c. 120°
d. 130°
e. 150°  *(circled)*

*handwritten:* $x + 5x = 180$
$6x = 180$
$x = 15$

**3.** If the measure of angle 4 is $6x + 20$ and the measure of angle 7 is $10x - 40$, what is the measure of angle 6?
a. 60°
b. 70°
c. 110°
d. 116°
e. 120°

*handwritten:* $6x + 20 + 10x - 40 = 180$
$16x - 20 = 180$
$16x = 20$

**4.** Which of the following is NOT true if the measure of angle 3 is 90°?
a. Angles 1 and 2 are complementary.
b. Angles 3 and 6 are supplementary.
c. Angles 5 and 7 are adjacent.
d. Angles 5 and 7 are congruent.
e. Angles 4 and 8 are supplementary and congruent.  *(circled)*

**5.** If the measure of angle 2 is $8x + 10$ and the measure of angle 6 is $x^2 - 38$, what is the measure of angle 8?
a. 42°
b. 74°
c. 84°
d. 108°
e. 138°

*handwritten:* $8x + 10 = x^2 - 38$
$8x + 48 = x^2$

Use the diagram below to answer questions 6–10. Lines *AE*, *BF*, *GD*, and ray *OC* all intersect at point *O* (not labeled), and line *AE* is perpendicular to ray *OC*.

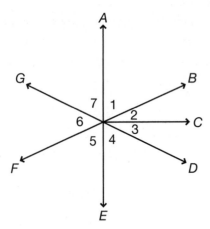

**6.** Which of the following pairs of angles are complementary?
   **a.** angles 1 and 2
   **b.** angles 2 and 3
   **c.** angles 1 and 4
   **d.** angles 1, 2, and 3
   **e.** angles 1, 6, and 7

**7.** Which of the following number sentences is NOT true?
   **a.** angle 1 + angle 2 = angle 3 + angle 4
   **b.** angle 1 + angle 2 + angle 3 + angle 7 = angle 4 + angle 5 + angle 6
   **c.** angle 2 + angle 3 = angle 6
   **d.** angle 1 + angle 7 = angle 2 + angle 3
   **e.** angle 2 + angle 3 + angle 4 + angle 5 = 180

**8.** If the measure of angle 3 is $2x + 2$ and the measure of angle 4 is $5x - 10$, what is the measure of angle 7?
   **a.** 14°
   **b.** 30°
   **c.** 60°
   **d.** 90°
   **e.** 115°

**9.** If angle 1 measures 62° and angle 4 measures 57°, what is the measure of angle 6?
   **a.** 33°
   **b.** 61°
   **c.** 72°
   **d.** 95°
   **e.** 119°

**10.** If the measure of angle 3 is $5x + 3$ and the measure of angle 4 is $15x + 7$, what is the sum of angles 5 and 6?
   **a.** 67°
   **b.** 113°
   **c.** 134°
   **d.** 157°
   **e.** cannot be determined

Use the diagram below to answer questions 11–15. Lines *EF* and *GH* are parallel to each other and line *IJ* is perpendicular to lines *EF* and *GH*.

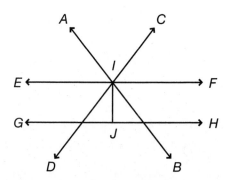

**11.** Which of the following pairs of angles must be congruent?
a. angles *AIC* and *KIL*
b. angles *KIJ* and *JIL*
c. angles *AIE* and *JIL*
d. angles *AIF* and *KIF*
e. angles *ILH* and *IKG*

**12.** If the measure of angle *JLI* is $8x - 4$ and the measure of angle *JIL* is $5x + 3$, what is the measure of angle *AIE*?
a. 14°
b. 38°
c. 48°
d. 52°
e. 128°

**13.** If the measure of angle *GKI* is $15x - 4$ and the measure of angle *CIF* is $x^2$, what is the measure of angle *EIC*?
a. 64°
b. 116°
c. 120°
d. 124°
e. 144°

**14.** If angles *LIF* and *EIK* are congruent, and the measure of angle *LIF* is 6 greater than the measure of angle *JIL*, what is the measure of angle *KIJ*?
a. 42°
b. 48°
c. 54°
d. 87°
e. 90°

**15.** If angle *JLB* is three and a half times the size of angle *LIF*, what is the measure of angle *EIL*?
a. 120°
b. 125°
c. 130°
d. 135°
e. 140°

Use the diagram below to answer questions 16–20. Lines *CD* and *GH* are perpendicular to each other, and every line goes through point *O* (not labeled). Every angle is greater than 0°.

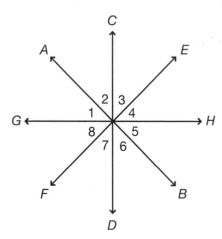

**16.** If lines *EF* and *AB* are perpendicular, the measure of angle 4 is $11x$, and the measure of angle 1 is $7x$, what is the measure of angle 4?

   **a.** 8°
   **b.** 35°
   **c.** 44°
   **d.** 45°
   **e.** 55°

**17.** Which of the following number sentences is true?

   **a.** angle 2 + angle 3 + angle 4 = angle 5 + angle 6 + angle 7
   **b.** angle 3 + angle 4 = angle 6 + angle 7
   **c.** angle 1 + angle 2 + angle 3 = angle 2 + angle 3 + angle 4
   **d.** angle 2 + angle 6 = angle 3 + angle 7
   **e.** angle 4 + angle 5 + angle 6 = angle 1 + angle 2 + angle 8

**18.** If the sum of angles 2 and 3 is $x^2$ and the sum of angles 6 and 7 is $10x$, what is the sum of angles 4 and 5?

   **a.** 80°
   **b.** 88°
   **c.** 90°
   **d.** 98°
   **e.** 100°

**19.** If the measure of angle 1 is $3x + 5$ and the measure of angle 6 is $5x - 3$, what is the measure of angle 2?

   **a.** 17°
   **b.** 38°
   **c.** 52°
   **d.** 73°
   **e.** cannot be determined

**20.** If angle 4 is congruent to angle 7, which of the following is NOT true?

   **a.** angle 4 = angle 3
   **b.** angle 2 = angle 3
   **c.** angle 3 = angle 8
   **d.** angle 7 = angle 3
   **e.** angle 7 = angle 8

Use the diagram below to answer questions 21–25. Lines *EF* and *GH* are parallel to each other, lines *CD* and *IJ* are parallel to each other, and line *AB* is perpendicular to lines *EF* and *GH*. Every numbered angle is greater than 0°.

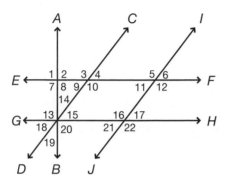

**21.** If the sum of the measures of angles 8 and 9 is 133°, what is the measure of angle 3?

   **a.** 43°

   **b.** 47°

   **c.** 90°

   **d.** 133°

   **e.** 137°

**22.** If the sum of the measures of angles 21 and 5 is $x^2 + 11$ and the measure of angle 3 is $9x + 1$, what is the measure of angle 10?

   **a.** 24°

   **b.** 62°

   **c.** 100°

   **d.** 118°

   **e.** 156°

**23.** Which of the following must be true?

   **a.** angle 13 + angle 14 = angle 13 + angle 15

   **b.** angle 16 + angle 10 = angle 11 + angle 15

   **c.** angle 1 + angle 8 = angle 19 + angle 15 + angle 20

   **d.** angle 8 + angle 9 = 180

   **e.** angle 14 + angle 15 + angle 16 = 180

**24.** If angle 5 = $8x - 4$ and angle 22 = $7x + 11$, what is the measure of angle 16?

   **a.** 92°

   **b.** 95°

   **c.** 100°

   **d.** 116°

   **e.** 124°

**25.** If the sum of the measures of angles 20 and 14 is $15x + 6$, and the sum of the measures of angles 20 and 15 is $18x$, what is the measure of angle 19?

   **a.** 18°

   **b.** 36°

   **c.** 45°

   **d.** 54°

   **e.** 90°

# 10 ▶ Triangles

## ▶ Interior Angles

The three **interior angles** of a triangle add up to 180°. Triangles are named by their vertices. The triangle below is named *ABC*, since the vertices *A*, *B*, and *C* comprise the triangle. This triangle could also be named *BCA*, *BAC*, *ACB*, *CBA*, or *CAB*. The vertices must be named in order, starting from any one of the vertices.

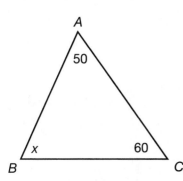

If you know the measure of two angles of a triangle, you can find the measure of the third angle by adding the measures of the first two angles and subtracting that sum from 180. The third angle of triangle *ABC* at left is equal to $180 - (50 + 60) = 180 - 110 = 70°$.

## ► Exterior Angles

The **exterior angles** of a triangle are the angles that are formed outside the triangle. Here is another look at triangle *ABC*, with each side of the triangle extended.

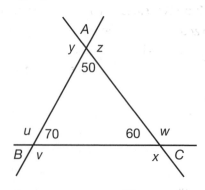

Adjacent interior and exterior angles are supplementary. Angle *y* and the angle that measures 50° are supplementary. Angle *z* is also supplementary to the angle that measures 50°, since these angles form a line. The measure of angle *y* is equal to 180 − 50 = 130°. Since angle *z* is also supplementary to the 50° angle, angle *z* also measures 130°. Notice that angles *y* and *z* are vertical angles—another reason these two angles are equal in measure.

The measure of an exterior angle is equal to the sum of the two interior angles to which the exterior angle is not adjacent. You already know angle *y* measures 130°, since it and angle *BAC* are supplementary. However, you could also find the measure of angle *y* by adding the measures of the other two interior angles. Angle *ABC*, 70, plus angle *ACB*, 60, is equal to the measure of the exterior angle of *BAC*: 70 + 60 = 130°, the measure of angles *y* and *z*.

If you find the measure of one exterior angle at each vertex, the sum of these three exterior angles is 360°. The measure of angle *y* is 130°. The measure of angle *u* is 110°, since it is supplementary to the 70° angle (180 − 70 = 110) and because the sum of the other interior angles is 110° (50 + 60 = 110). The measure of angle *w* is 120°, since it is supplementary to the 60° angle (180 − 60 = 120) and because the sum of the other interior angles is 120° (70 + 50 = 120). The sum of angles *y*, *u*, and *w*, is 130 + 110 + 120 = 360°.

## ► Types of Triangles

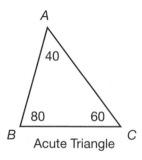

Acute Triangle

If the measure of the largest angle of a triangle is less than 90°, the triangle is an **acute triangle**. The largest angle in triangle *ABC* measures 80°; therefore, it is an acute triangle.

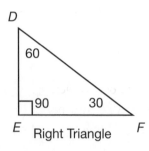

Right Triangle

If the measure of the largest angle of a triangle is equal to 90°, the triangle is a **right triangle**. The largest angle in triangle *DEF* measures 90°; therefore, it is a right triangle.

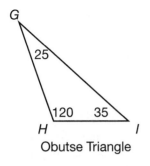

Obutse Triangle

If the measure of the largest angle of a triangle is greater than 90°, the triangle is an **obtuse triangle**. The largest angle in triangle *GHI* measures 120°; therefore, it is an obtuse triangle.

There are three other types of triangles. If no two sides or angles of a triangle are equal, the triangle is **scalene**. If exactly two sides (and therefore, two angles) of a triangle are equal, the triangle is **isosceles**. If all three sides (and therefore, all three angles) of a triangle are equal, the triangle is **equilateral**.

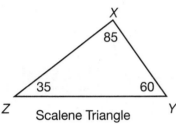

Scalene Triangle

In a triangle, the side opposite the largest angle of the triangle is the longest side, and the side opposite the smallest angle is the shortest side. In scalene triangle *XYZ*, side *YZ* is the longest side, since it is the side opposite the largest angle. In an isosceles triangle, the sides opposite the equal angles are the equal sides. In a right triangle, the angle opposite the right angle is the **hypotenuse**, which is always the longest side of the triangle.

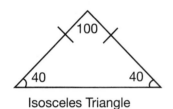

Isosceles Triangle

## ▶ Similar Triangles

Two triangles are **similar** if the measures of their corresponding angles are identical. The lengths of the corresponding sides of similar triangles can be different—it is the measures of the angles of the triangles that make the triangles similar.

The sides of similar triangles can be expressed with a ratio. If triangles *JKL* and *MNO* are similar, and each side of triangle *JKL* is twice as long as its corresponding side of triangle *MNO*, the ratio of the sides of triangle *JKL* to the sides of triangle *MNO* is 2:1.

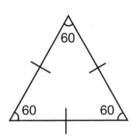

Equilateral Triangle

## ▶ Practice

**1.** If the measure of angle *A* of triangle *ABC* is 3*x*, the measure of angle *B* is 5*x*, and the measure of angle *C* is 4*x*, what is the value of *x*?

   a. 12°
   b. 15°
   c. 20°
   d. 30°
   e. 45°

**2.** If the measure of angle *A* of triangle *ABC* is 5*x* + 10, the measure of angle *B* is *x* + 10, and the measure of angle *C* is 2*x*, which of the following is true of triangle *ABC*?

   a. Triangle *ABC* is acute and scalene.
   b. Triangle *ABC* is acute but not scalene.
   c. Triangle *ABC* is right but not isosceles.
   d. Triangle *ABC* is obtuse but not scalene.
   e. Triangle *ABC* is obtuse and scalene.

**3.** The measure of an angle exterior to angle $F$ of triangle $DEF$ measures $16x + 12$. If angle $F$ measures $8x$, what is the measure of angle $F$?

a. 7°

b. 49°

c. 56°

d. 102°

e. 124°

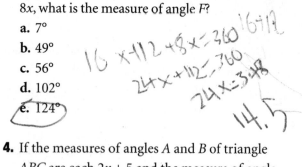

**4.** If the measures of angles $A$ and $B$ of triangle $ABC$ are each $2x + 5$ and the measure of angle $C$ is $3x - 5$, what is the measure of angle exterior to angle $A$?

a. 25°

b. 55°

c. 70°

d. 110°

e. 125°

$$7x + 5 = 180$$
$$7x = 175$$
$$x = 25$$

55

**5.** The measure of an angle exterior to angle $F$ of triangle $DEF$ measures 120°. Which of the following must be true?

a. Triangle $DEF$ is obtuse.

b. Triangle $DEF$ is acute.

c. Triangle $DEF$ is equilateral.

d. Triangle $DEF$ is scalene.

e. Triangle $DEF$ is not isosceles.

Use the diagram below to answer questions 6–9. The diagram is not to scale, and every angle is greater than 0°.

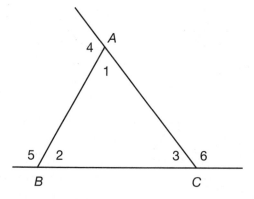

**6.** Which of the following number sentences is NOT true?

a. angle 1 + angle 2 = angle 5

b. angle 4 + angle 1 = angle 3 + angle 6

c. angle 3 + angle 2 = angle 4

d. angle 1 + angle 2 + angle 3 = 180

e. angle 4 + angle 5 + angle 6 = 360

**7.** If angle 6 measures 115° and angle 2 measures 75°, what is the measure of angle 4?

a. 75°

b. 105°

c. 115°

d. 140°

e. 190°

**8.** If angle 4 measures $7x + 2$, angle 5 measures $8x$, and angle 6 measures $8x - 10$, what is the measure of angle 1?
   **a.** 52°
   **b.** 62°
   **c.** 66°
   **d.** 114°
   **e.** 118°

**9.** If the measure of angle 1 is $x^2 + 1$, the measure of angle 2 is $9x - 7$, and the measure of angle 3 is $6x + 2$, which of the following is true?
   **a.** Triangle $ABC$ is isosceles.
   **b.** Triangle $ABC$ is obtuse.
   **c.** Triangle $ABC$ is equilateral.
   **d.** Triangle $ABC$ is scalene.
   **e.** Triangle $ABC$ is right.

**10.** If the measure of angle $F$ in triangle $DEF$ is half the sum of the measures of angles $D$ and $E$, what is the measure of an angle exterior to $F$?
   **a.** 30°
   **b.** 60°
   **c.** 90°
   **d.** 120°
   **e.** cannot be determined

Use the diagram below to answer questions 11–13. Triangles $ABC$ and $DEF$ are similar. The diagram is not to scale, and every angle is greater than 0°.

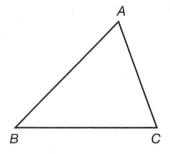
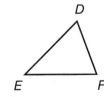

**11.** If the length of $\overline{AB}$ is 90, the length of $\overline{DE}$ is 60, and the length of $\overline{AC}$ is 72, what is the length of $\overline{DF}$?
   **a.** 30
   **b.** 36
   **c.** 42
   **d.** 48
   **e.** 108

**12.** Every side of triangle $DEF$ is greater than 3. If the length of $\overline{AB}$ is $10x - 2$, the length of $\overline{AC}$ is $6x$, the length of $\overline{DE}$ is $2x + 2$, and the length of $\overline{DF}$ is $x + 2$, what is the length of $\overline{DF}$?
   **a.** 3
   **b.** 4
   **c.** 6
   **d.** 10
   **e.** 12

**13.** If the length of $\overline{BC}$ is $x$ and the length of $\overline{EF}$ is $\frac{1}{5}x$, which of the following is NOT true?
   **a.** The length of $\overline{AB}$ is 5 times the length of $\overline{DE}$.
   **b.** The measure of angle $A$ is five times the measure of angle $D$.
   **c.** The length of $\overline{DF}$ is $\frac{1}{5}$ the length of $\overline{AC}$.
   **d.** Angle $B$ is congruent to angle $E$.
   **e.** The sum of the sides of triangle $ABC$ is five times the sum of the sides of triangle $DEF$.

**14.** Triangles $JKL$ and $MNO$ are congruent and equilateral. If side $JK$ measures $6x + 3$ and side $MN$ measures $x^2 - 4$, what is the length of side $NO$?
   **a.** 7
   **b.** 42
   **c.** 45
   **d.** 60
   **e.** cannot be determined

**15.** Triangles *GHI* and *PQR* are similar and isosceles. If side *HG* of triangle *GHI* is congruent to side *PQ*, its corresponding side in triangle *PQR*, which of the following must be true?
a. Side *GI* is congruent to its corresponding side.
b. Angle *G* is congruent to angle *R*.
c. Angle *G* is congruent to angle *I*.
d. Side *GH* is congruent to side *PR*.
e. Angle *H* is not congruent to angle *R*.

Use the diagram below to answer questions 16–20. Lines *EF* and *GH* are parallel. The diagram is not to scale, and every angle is greater than 0°.

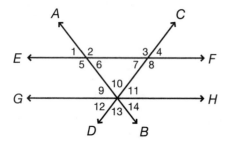

**16.** If angles 6 and 7 are congruent, and the measure of angle 13 is 94°, what is the measure of angle 8?
a. 43°
b. 86°
c. 94°
d. 137°
e. 147°

**17.** If angle 3 measures $10x + 15$, angle 10 measures $8x - 3$, and angle 6 measures $3x + 7$, what is the measure of angle 7?
a. 40°
b. 55°
c. 65°
d. 85°
e. 125°

**18.** The measure of angle 5 is $10x + 2$, the measure of angle 11 is $4x - 4$, and the measure of angle 13 is $7x - 6$. What is the sum of angles 10 and 7?
a. 44°
b. 58°
c. 102°
d. 122°
e. 136°

**19.** The measure of angle 11 is twice the measure of angle 14, and the measure of angle 8 is 2.5 times the measure of angle 14. What is the measure of angle 10?
a. 20°
b. 40°
c. 60°
d. 80°
e. 100°

**20.** If the triangle formed by lines *AB*, *CD*, and *GH* is an isosceles right triangle, angle 8 is greater than angle 7, and angle 8 is congruent to angle 5, which of the following is NOT true?
a. angle 2 = angle 9 + angle 12
b. angle 13 = 90°
c. Angles 11 and 14 are complementary.
d. angle 1 = angle 4
e. angle 3 = angle 12 + angle 13

# 11 ▶ Right Triangles

## ▶ Pythagorean Theorem

As you saw in the previous chapter, right triangles are triangles in which one angle measures 90°. Given the lengths of two sides of a right triangle, the Pythagorean theorem can be used to find the missing side of the right triangle.

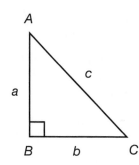

The **Pythagorean theorem** states that the sum of the squares of the bases of a triangle is equal to the square of the hypotenuse of the triangle. If $a$ is the length of one base, $b$ is the length of the other base, and $c$ is the length of the hypotenuse, then $a^2 + b^2 = c^2$.

The formula can be rewritten to find any of the three sides. For instance, $a^2 = c^2 - b^2$ and $b^2 = c^2 - a^2$.

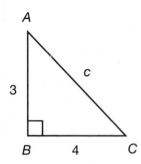

In triangle $ABC$ at left, the bases are sides $AB$ and $BC$, since the hypotenuse, $\overline{AC}$, is the side opposite the right angle. To find $\overline{AC}$, use the Pythagorean theorem. Square sides $a$ and $b$, and then add those squares; $a^2 + b^2 = c^2$, $(3)^2 + (4)^2 = c^2$, $9 + 16 = c^2$, $25 = c^2$. Now, take the square root of both sides of the equation to find $c$; $c = \sqrt{25}$, $c = 5$. The length of side $AC$ is 5 units.

## ▶ Common Pythagorean Triples

The three lengths of a right triangle are often referred to as **Pythagorean triples.** You just saw that 3:4:5 is a Pythagorean triple. Another common Pythagorean triple is 5:12:13. Multiples of these triples are also Pythagorean triples; 6:8:10, 9:12:15, and 12:16:20 are all multiples of the 3:4:5 triple. Multiples of the 5:12:13 triple are 10:24:26 and 15:36:39.

**Common Right Triangles**

There are two common types of right triangles often seen on the SAT: the 45-45-90 right triangle and the 30-60-90 right triangle.

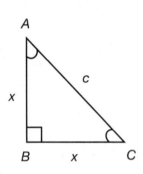

The **45-45-90 right triangle** is an isosceles right triangle. Its angles measure 45°, 45°, and 90°. The bases of this right triangle are equal. If the bases each measure 2 units, what is the length of the hypotenuse? Using the Pythagorean theorem, you can find that $2^2 + 2^2 = c^2$, $8 = c^2$, and $c = \sqrt{8}$, or $2\sqrt{2}$. What if each base measured 3 units? Then, $3^2 + 3^2 = c^2$, $18 = c^2$, and $c = \sqrt{18}$, or $3\sqrt{2}$. Notice that in both cases, the length of the hypotenuse is $\sqrt{2}$ times the length of a base of the triangle. This is always the case with 45-45-90 right triangles. The length of the hypotenuse is $\sqrt{2}$ times the length of a base. In the same way, the length of a base is $\frac{\sqrt{2}}{2}$ times the length of the hypotenuse.

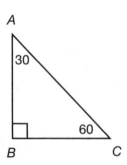

The **30-60-90 right triangle** is another type of right triangle. Its angles measure 30°, 60°, and 90°. If side $BC$ of triangle $ABC$ at right is 2 units and side $AB$ is $2\sqrt{3}$ units, what is the length of the hypotenuse? Using the Pythagorean theorem again, you can find that $2^2 + (2\sqrt{3})^2 = c^2$, $4 + 12 = c^2$, $c^2 = 16$, and $c = 4$. In every 30-60-90 right triangle, the length of the hypotenuse is 2 times the length of the shorter base (the base that is opposite the 30° angle). The length of the longer base (the base opposite the 60° angle) is always $\sqrt{3}$ times the length of the shorter base.

# ▶ Trigonometry

There are three trigonometry functions that will appear on the SAT: sine, cosine, and tangent.

The **sine** of an angle is equal to the length of the base opposite the angle divided by the length of the hypotenuse.

The **cosine** of an angle is equal to the length of the base adjacent to the angle divided by the length of the hypotenuse.

The **tangent** of an angle is equal to the length of the base opposite the angle divided by the length of the base adjacent to the angle.

Use the mnemonic device **SOHCATOA**: $\text{Sine} = \frac{\text{Opposite}}{\text{Hypotenuse}}$, $\text{Cosine} = \frac{\text{Adjacent}}{\text{Hypotenuse}}$, and $\text{Tangent} = \frac{\text{Opposite}}{\text{Adjacent}}$.

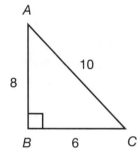

In triangle $ABC$, what are the sine, cosine, and tangent of angle $C$?

First, find the sine. Since $\overline{AB}$ is the side opposite angle $C$, divide the length of $\overline{AB}$ by the length of the hypotenuse, $\overline{AC}$: $\frac{8}{10}$, or $\frac{4}{5}$, is the sine of angle $C$. The cosine of $C$ is equal to the side adjacent to angle $C$, side $BC$, divided by $\overline{AC}$: $\frac{6}{10}$, or $\frac{3}{5}$. The tangent of $C$ is equal to $\frac{\overline{AB}}{\overline{BC}} = \frac{8}{6}$, or $\frac{4}{3}$.

Be sure to know the sine, cosine, and tangent of these common angles:

|         | 30°                   | 45°                     | 60°                   |
|---------|-----------------------|-------------------------|-----------------------|
| sine    | $\frac{1}{2}$         | $\frac{\sqrt{2}}{2}$    | $\frac{\sqrt{3}}{2}$  |
| cosine  | $\frac{\sqrt{3}}{2}$  | $\frac{\sqrt{2}}{2}$    | $\frac{1}{2}$         |
| tangent | $\frac{\sqrt{3}}{3}$  | $1$                     | $\sqrt{3}$            |

# ▶ Practice

**1.** The bases of a right triangle measure $x - 3$ and $x + 4$. If the hypotenuse of the triangle is $2x - 3$, what is the length of the hypotenuse?
   a. 4
   b. 5
   c. 8
   d. 12
   e. 13

**2.** If the length of a base of right triangle $DEF$ is 8 units and the hypotenuse of triangle $DEF$ is $8\sqrt{5}$ units, what is the length of the other base?
   a. 4 units
   b. 8 units
   c. $8\sqrt{2}$ units
   d. 16 units
   e. 32 units

**3.** If the lengths of the bases of right triangle $GHI$ are 9 units and 15 units respectively, what is length of the hypotenuse of triangle $GHI$?
a. $6\sqrt{2}$ units
b. $3\sqrt{34}$ units
c. $9\sqrt{34}$ units
d. 18 units
e. 30 units

**4.** If the longer base of triangle $XYZ$ is three times the length of the shorter base, $a$, what is the length of the hypotenuse in terms of the shorter base?
a. $a\sqrt{2}$
b. $a\sqrt{3}$
c. $a\sqrt{10}$
d. $3a$
e. $a^2$

**5.** Which of the following triangles is a multiple of the triangle with sides measuring $x$, $x - 5$, and $x + 5$, when $x$ is greater than 0?
a. a triangle with sides measuring 1, 2, and 3
b. a triangle with sides measuring 3, 4, and 5
c. a triangle with sides measuring 1, 5, and 10
d. a triangle with sides measuring 5, 10, and 15
e. a triangle with sides measuring 5, 10, and 20

**6.** Angle $T$ of right triangle $TUV$ measures 45°. If base $TU$ measures 10 units, what is the length of the hypotenuse of triangle $TUV$?
a. $\sqrt{2}$ units
b. $10\sqrt{2}$ units
c. $20\sqrt{2}$ units
d. $45\sqrt{2}$ units
e. 100 units

**7.** The hypotenuse of an isosceles right triangle measures $x$ units. What is the length of a base of the triangle?
a. $\frac{x\sqrt{2}}{2}$ units
b. $\frac{\sqrt{x}}{2}$ units
c. $\frac{1}{2}x$ units
d. $x\sqrt{2}$ units
e. $x^2$ units

**8.** Angle $Q$ of triangle $PQR$ is a right angle, and side $QR$ measures 4 units. If the sine of angle $P$ is equal to the sine of angle $R$, what is the length of side $PR$?
a. $\frac{\sqrt{2}}{4}$ units
b. $\frac{\sqrt{2}}{2}$ units
c. $2\sqrt{2}$ units
d. $4\sqrt{2}$ units
e. cannot be determined

Use the diagram below to answer questions 9–10. Line $AB$ is parallel to line $CD$, and line $GH$ is perpendicular to lines $AB$ and $CD$. The diagram is not to scale.

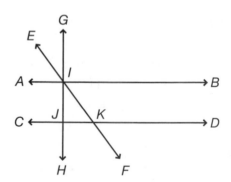

**9.** If line segment $JK$ is 16 units long, what is the length of line segment $IK$?
a. 4 units
b. $4\sqrt{2}$ units
c. $8\sqrt{2}$ units
d. $16\sqrt{2}$ units
e. cannot be determined

**10.** If line *CD* is 25 units from line *AB*, and the measure of angle *IKD* is 135°, what is the length of line segment *IK*?

   **a.** 5 units

   **b.** $5\sqrt{2}$ units

   **c.** $\frac{25\sqrt{2}}{2}$ units

   **d.** $25\sqrt{2}$ units

   **e.** cannot be determined

**11.** Given right triangle *ABC* with right angle *B*, angle *A* is twice the size of angle *C*. If the measure of side *AB* is 7 units, what is the measure of side *AC*?

   **a.** $7\sqrt{2}$ units

   **b.** $7\sqrt{3}$ units

   **c.** $\sqrt{7}$ units

   **d.** $\sqrt{14}$ units

   **e.** 14 units

Use the diagram below to answer questions 12–14. Angle *C* measures 60°. The diagram is not to scale.

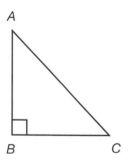

**12.** If the length of side *AB* is 9 cm, what is the length of side *BC*?

   **a.** $\frac{\sqrt{3}}{3}$ units

   **b.** 3 units

   **c.** 4.5 units

   **d.** $3\sqrt{3}$ units

   **e.** $9\sqrt{3}$ units

**13.** If the hypotenuse of triangle *ABC* is $6x + 2$ units long, what is the length of side *AB*?

   **a.** $\sqrt{6x + 2}$ units

   **b.** $3x + 1$ units

   **c.** $(3x + 1)\sqrt{3}$ units

   **d.** $(6x + 2)\sqrt{2}$ units

   **e.** $12x + 4$ units

**14.** If the sum of sides *BC* and *AC* is 12 units, what is the length of side *AB*?

   **a.** $2\sqrt{3}$ units

   **b.** 4 units

   **c.** $4\sqrt{3}$ units

   **d.** 8 units

   **e.** $12\sqrt{3}$ units

**15.** Triangle *ABC* is an equilateral triangle and triangle *CDE* is a right triangle. If the length of side *AE* is 20 units, what is the length of side *BD*?

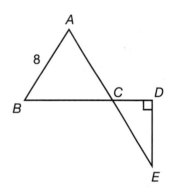

   **a.** 12 units

   **b.** 14 units

   **c.** 16 units

   **d.** 18 units

   **e.** 20 units

**16.** If angle $B$ of isosceles triangle $ABC$ is a right angle, what is the tangent of angle $A$?

   **a.** 0

   **b.** $\frac{1}{2}$

   **c.** 1

   **d.** $\frac{\sqrt{3}}{3}$

   **e.** cannot be determined

Use the diagram below to answer questions 17–20. The diagram is not to scale.

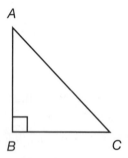

**17.** If line $AB$ measures $3x - 6$, line $BC$ measures $x^2 - 2x$, and line $AC$ measures $2x + 2$, what is the tangent of angle $A$?

   **a.** $\frac{x}{3}$

   **b.** $\frac{3}{x}$

   **c.** $\frac{3x - 6}{2x + 2}$

   **d.** $\frac{(x^2 - 2x)}{(2x + 2)}$

   **e.** $\frac{2x + 2}{3x - 6}$

**18.** If the cosine of angle $C$ is $\frac{x}{y}$, triangle $ABC$ is not isosceles, and $x$ does not equal $y$, which of the following is also equal to $\frac{x}{y}$?

   **a.** sine of angle $A$

   **b.** sine of angle $C$

   **c.** cosine of angle $A$

   **d.** tangent of angle $A$

   **e.** 2

**19.** If the sine of angle $A$ is $\frac{15}{17}$, what is the cosine of angle $A$?

   **a.** $\frac{8}{17}$

   **b.** $\frac{8}{15}$

   **c.** $\frac{12}{17}$

   **d.** $\frac{17}{15}$

   **e.** $\frac{15}{8}$

**20.** If the tangent of angle $A$ is 0.75 and the measure of side $AB$ is 4 less than the measure of side $AC$, what is the length of side $BC$?

   **a.** 3 units

   **b.** 4 units

   **c.** 6 units

   **d.** 8 units

   **e.** 12 units

Use the diagram below to answer questions 21–25. Line *EF* is parallel to line *GH*, and line *JL* is perpendicular to lines *EF* and *GH*. The diagram is not to scale.

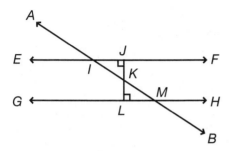

**21.** If the length of side *KM* is 10 units and the length of side *LM* is 5 units, what is the measure of angle *KIJ*?

a. 0.5°

b. 30°

c. 45°

d. 50°

e. 60°

**22.** If the length of side *LM* is 8 units and the sum of angles *IKJ* and *LKM* is 60°, what is the length of side *KM*?

a. $\frac{16\sqrt{3}}{3}$ units

b. $8\sqrt{2}$ units

c. $8\sqrt{3}$ units

d. 16 units

e. $16\sqrt{3}$ units

**23.** If the tangent of angle *JIK* is $\sqrt{3}$, then

a. side *LK* is $\sqrt{3}$ times the length of side *LM*.

b. side *IJ* is $\sqrt{3}$ times the length of side *JK*.

c. the sine of triangle *JIK* is $\frac{\sqrt{2}}{2}$.

d. the length of side *IK* is twice the length of side *JK*.

e. the length of side *IJ* is equal to the length of side *JK*.

**24.** If angle *KMH* is three times the size of angle *KML* and the length of side *JK* is $x\sqrt{6}$ units, what is the length of side *IK*?

a. $x\sqrt{2}$ units

b. $2x\sqrt{3}$ units

c. $3x$ units

d. $3x\sqrt{2}$ units

e. $2x\sqrt{6}$ units

**25.** If the length of side *IJ* is $2x - 2$, the length of side *IK* is $2x + 1$, the length of side *KM* is $3x - 1$, and the length of side *LM* is $2x + 2$, what is the length of side *LM*?

a. 9 units

b. 12 units

c. 15 units

d. 16 units

e. 20 units

# 12 ▶ Polygons

## ▶ Types of Polygons

A **polygon** is a closed figure with three or more sides. For example, a triangle is a polygon; so is a square. A circle, however, is not—it is a closed figure, but it does not have at least three sides.

You should be familiar with the names of polygons that have three to ten sides. A **triangle** is a polygon with three sides. A **quadrilateral** has four sides, a **pentagon** has five sides, a **hexagon** has six sides, a **heptagon** has seven sides, an **octagon** has eight sides, a **nonagon** has nine sides, and a **decagon** has ten sides.

## ▶ Regular Polygons

A **regular polygon** is an equilateral polygon—all sides of the polygon are the same length. An equilateral triangle is a regular polygon.

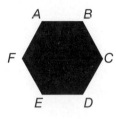

If $A$ is the sum of the interior angles of a polygon, and $s$ is the number of sides of the polygon, then $A = 180(s - 2)$.

Regular hexagon *ABCDEF* has six sides. Therefore, the sum of its **interior angles** is $180(6 - 2) = 180(4) = 720°$. Since hexagon *ABCDEF* is a regular polygon, every side is congruent to every other side, and every angle is congruent to every other angle. To find the measure of an angle of a regular polygon, divide the sum of the interior angles by the number of sides (angles) of the polygon. Every angle of a regular hexagon measures $\frac{720}{6} = 120°$.

## ▶ Exterior Angles

The sum of the **exterior angles** of any polygon is 360°. Even if the polygon has 100 sides, its exterior angles will still add up to 360°.

## ▶ Similarity

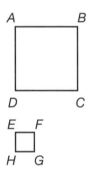

Two polygons are **similar** if the measures of their corresponding angles are identical. As with similar triangles, the lengths of the corresponding sides of similar polygons can be different—it is the measures of the angles of the polygons that make the polygons similar.

The sides of similar polygons can be expressed with a ratio. If polygons *ABCD* and *EFGH* are similar, and each side of polygon *ABCD* is four times the length of its corresponding side of polygon *EFGH*, the ratio of the sides of polygon *ABCD* to the sides of polygon *EFGH* is 4:1. The ratio of the perimeter of polygon *ABCD* to the perimeter of polygon *EFGH* is also 4:1.

## ▶ Perimeter

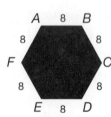

The **perimeter** of a polygon is the sum of the lengths of its sides. To find the perimeter of a triangle, add the lengths of each of its 3 sides. To find the length of a polygon with 12 sides, add the lengths of all 12 sides. The perimeter of hexagon *ABCDEF* is 48, since $8 + 8 + 8 + 8 + 8 + 8 = 48$.

The perimeter of hexagon *ABCDEF* is also equal to $(6)(8)$. The perimeter of a regular polygon is equal to the length of one side of the polygon multiplied by the number of sides of the polygon.

# ▶ Practice

1. Heather draws two regular pentagons. Which of the following is NOT always true?
   a. The two regular pentagons are similar.
   b. The two regular pentagons are congruent.
   c. The sum of the interior angles of each pentagon is 540°.
   d. The sum of the exterior angles of each pentagon is 360°.
   e. Each side of each pentagon is congruent to every other side of that pentagon.

2. The sum of the interior angles of a polygon is equal to three times the sum of its exterior angles. How many sides does the polygon have?
   a. 6 sides
   b. 8 sides
   c. 10 sides
   d. 12 sides
   e. cannot be determined

3. Andrea draws a polygon with $x$ number of sides. The sum of the interior angles of her polygon is 60 times its number of sides. How many sides does Andrea's polygon have?
   a. 3 sides
   b. 4 sides
   c. 5 sides
   d. 6 sides
   e. 10 sides

4. If the sum of the interior angles of a polygon is equal to the sum of the exterior angles, which of the following statements must be true?
   a. The polygon is a regular polygon.
   b. The polygon has 2 sides.
   c. The polygon has 4 sides.
   d. The polygon has 6 sides.
   e. It is impossible for these sums to be equal.

5. The sum of the interior angles of a polygon is $9x^2$. If $x$ is 3 greater than the number of sides of the polygon, how many sides does the polygon have?
   a. 6 sides
   b. 7 sides
   c. 10 sides
   d. 12 sides
   e. 13 sides

6. Pentagons $ABCDE$ and $FGHIJ$ are similar. The ratio of each side of pentagon $ABCDE$ to its corresponding side of pentagon $FGHIJ$ is 4:1. If $\overline{AB}$ and $\overline{FG}$ are corresponding sides, and the length of $\overline{AB}$ is $4x + 4$, what is the length of $\overline{FG}$?
   a. $x + 1$
   b. $2x + 2$
   c. $4x + 4$
   d. $16x + 16$
   e. $20x + 20$

7. Quadrilaterals $ABCD$ and $EFGH$ are similar. If the perimeter of quadrilateral $ABCD$ is equal to $4y^2$, what is the perimeter of quadrilateral $EFGH$?
   a. $y^2$
   b. $2y$
   c. $4y^2$
   d. $16y^4$
   e. cannot be determined

8. If the ratio of the perimeter of octagon $ABCDEFGH$ to the perimeter of octagon $STUVWXYZ$ is 1:1, which of the following must be true?
   a. The two octagons are regular octagons.
   b. The ratio of $\overline{AB}$ to $\overline{ST}$ is 1:1.
   c. At least one side of octagon $ABCDEFGH$ is equal to at least one side of octagon $STUVWXYZ$.
   d. The sums of the interior angles of each octagon are equal.
   e. The two octagons are similar, but not necessarily congruent.

**9.** Polygon *ABCDEF* is similar to polygon *GHIJKL*. If side *AB* is $12x + 6x$ and side *GH* is $8x + 4x$, and these sides are corresponding sides, what is the ratio of the perimeter of polygon *GHIJKL* to the perimeter of polygon *ABCDEF*?

a. 1:3
b. 2:1
c. 2:3
d. 3:1
e. 3:2

**10.** Regular polygon *ABCDE* is similar to regular polygon *TUVWX*. Side *AB* is $5x - 1$, and side *TU* is $4x - 2$. These sides are corresponding sides. If the ratio of the perimeter of polygon *ABCDE* to the perimeter of polygon *TUVWX* is 4:3, what is the perimeter of polygon *ABCDE*?

a. 20 units
b. 24 units
c. 54 units
d. 72 units
e. 120 units

**11.** If the perimeter of the figure below is 60 units, what is the length of $\overline{BC}$?

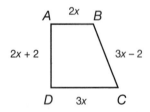

a. 6 units
b. 10 units
c. 14 units
d. 16 units
e. 18 units

**12.** The perimeter of a regular seven-sided polygon is $11x - 4$. If $x = 8$, what is the length of one side of the polygon?

a. 8 units
b. 12 units
c. 13 units
d. 15 units
e. 56 units

**13.** The ratio of the lengths of a side of regular pentagon *ABCDE* to the length of a side of regular hexagon *PQRSTU* is 5:6. If the perimeter of pentagon *ABCDE* is 75 units, what is the perimeter of hexagon *PQRSTU*?

a. 15 units
b. 18 units
c. 36 units
d. 90 units
e. 108 units

**14.** What is the perimeter of an isosceles right triangle whose hypotenuse is $5\sqrt{2}$ units?

a. 5 units
b. $15 + \sqrt{2}$ units
c. $10 + 5\sqrt{2}$ units
d. $15\sqrt{2}$ units
e. $25\sqrt{2}$ units

**15.** If the sum of the interior angles of a regular polygon equals 720°, and the length of one side of the polygon is $3x^2$, what is the perimeter of the polygon?

a. $18x^2$ units
b. $18x^{12}$ units
c. $24x^2$ units
d. $24x^{12}$ units
e. $27x^2$ units

# 13 ▶ Quadrilaterals

**A** **quadrilateral** is a polygon with four sides. The interior angles of any quadrilateral add up to 360°. Since the sum of the exterior angles of any polygon is 360°, the sum of the exterior angles of a quadrilateral is 360°.

## ▶ Diagonals

The lines that connect opposite angles of a quadrilateral are its **diagonals**. A diagonal of a quadrilateral cuts the quadrilateral into two triangles.

# ► Types of Quadrilaterals

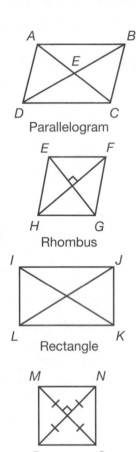

Parallelogram

Rhombus

Rectangle

Square

A **parallelogram** is a quadrilateral that has two pairs of parallel sides. The opposite sides of a parallelogram are parallel and congruent. Opposite angles are congruent and pairs of angles that are not opposite are supplementary. Sides *AB* and *CD* of parallelogram *ABCD* are parallel and congruent. Sides *BC* and *AD* are also parallel and congruent. Angles *A* and *C* are congruent, as are angles *B* and *D*. Angle *A* is supplementary to angles *B* and *C*; angle *D* is also supplementary to angles *B* and *C*. The diagonals of a parallelogram bisect each other.

Therefore, line segments *AE* = *EC* and line segments *BE* = *ED*. The two diagonals themselves, line segments *AC* and *BD*, are not necessarily congruent.

A **rhombus** is a parallelogram with four congruent sides. Since a rhombus is a parallelogram, its opposite sides are parallel and its opposite angles are congruent. Not only do the diagonals of a rhombus bisect each other, they are perpendicular to each other.

A **rectangle** is a parallelogram with four right angles. Since a rectangle is a parallelogram, its opposite sides are parallel and congruent. All four angles are congruent (all measure 90°) and the diagonals of a rectangle are congruent. In rectangle *IJKL* at left, side *IK* = side *JL*.

A **square** is a parallelogram with four congruent sides and four right angles. Also true is that a square is a rectangle with four congruent sides or that a square is a rhombus with four right angles. The diagonals of a square bisect each other, intersect at 90° angles, and are congruent.

# ► Practice

**1.** Andrew constructs a polygon with four sides and no right angles. His polygon
   **a.** could be a parallelogram, but cannot be a rectangle.
   **b.** could be a rhombus, but cannot be a parallelogram.
   **c.** could be a rectangle, but cannot be a square.
   **d.** could be a parallelogram or a rectangle, but not a square.
   **e.** cannot be a parallelogram, rhombus, rectangle, or square.

**2.** Angle *E* of rhombus *EFGH* measures 3*x* + 5. If the measure of angle *H* is 4*x*, what is the measure of angle *F*?
   **a.** 20°
   **b.** 80°
   **c.** 100°
   **d.** 120°
   **e.** 140°

**3.** DeDe has four identical line segments. Using them, she can form
   **a.** a square or a rhombus, but not a rectangle.
   **b.** a rhombus, but not a parallelogram.
   **c.** a square or a rectangle, but not a parallelogram.
   **d.** a square, but not a rhombus or a rectangle.
   **e.** a square, rhombus, rectangle, or parallelogram.

**4.** The diagonals of rectangle $ABCD$ intersect at $E$. Which of the following is NOT always true?
   **a.** angle $AEB$ = angle $DEC$
   **b.** angle $EDC$ = angle $EBA$
   **c.** angle $DAE$ = angle $EDA$
   **d.** angle $DCE$ = angle $ECB$
   **e.** angle $ECD$ = angle $ABE$

**5.** Monica draws a quadrilateral whose diagonals form four right triangles inside the quadrilateral. This quadrilateral
   **a.** must have four congruent angles.
   **b.** must have four congruent sides.
   **c.** must have congruent diagonals.
   **d.** must be a square.
   **e.** all of the above

**6.** The length of a rectangle is four less than twice its width. If $x$ is the width of the rectangle, what is the perimeter of the rectangle?
   **a.** $2x^2 - 4x$
   **b.** $3x - 4$
   **c.** $6x - 4$
   **d.** $6x + 4$
   **e.** $6x - 8$

**7.** The length of one side of a rhombus is $x^2 - 6$. If the perimeter of the rhombus is 168 units, what is the value of $x$?
   **a.** $4\sqrt{3}$
   **b.** 7
   **c.** 8
   **d.** $4\sqrt{6}$
   **e.** $\sqrt{174}$

**8.** The length of a rectangle is four times the length of a square. If the rectangle and the square share a side, and the perimeter of the square is 2 m, what is the perimeter of the rectangle?
   **a.** 5 units
   **b.** 6 units
   **c.** 8 units
   **d.** 10 units
   **e.** 20 units

**9.** If the perimeter of a parallelogram $ABCD$ is equal to the perimeter of rhombus $EFGH$ and the perimeter of square $IJKL$, then
   **a.** these three figures must be three congruent squares.
   **b.** every side of the rhombus must be congruent to every side of the square.
   **c.** the parallelogram must also be a rhombus.
   **d.** the parallelogram must not be a square or a rhombus.
   **e.** the rhombus could be a square, but the parallelogram must not be a square.

**10.** Four squares are joined together to form one large square. If the perimeter of one of the original squares was $8x$ units, what is the perimeter of the new, larger square?
   **a.** $16x$ units
   **b.** $20x$ units
   **c.** $24x$ units
   **d.** $32x$ units
   **e.** $64x$ units

**11.** What is the perimeter of a square with a diagonal measuring $2x\sqrt{2}$ units?
   **a.** $2x$ units
   **b.** $4x$ units
   **c.** $4x\sqrt{2}$ units
   **d.** $8x$ units
   **e.** $8x\sqrt{2}$ units

**12.** Diagonal *AC* of rectangle *ABCD* creates angle *ACB*, the tangent of which is 8. If the length of side *BC* is 8 units, what is the perimeter of rectangle *ABCD*?

   **a.** 18 units

   **b.** 32 units

   **c.** 64 units

   **d.** 72 units

   **e.** 144 units

**13.** If the perimeter of a square is equal to $5x + 1$ and the length of the diagonal of the square is $(2x - 2)\sqrt{2}$, what is the length of a side of the square?

   **a.** 2 units

   **b.** 4 units

   **c.** 8 units

   **d.** 10 units

   **e.** 11 units

**14.** Diagonal *AC* of rectangle *ABCD* creates angle *ACD*, the cosine of which is $\frac{12}{13}$. If the lengths of the sides of rectangle *ABCD* are all integers, which of the following could be the perimeter of rectangle *ABCD*?

   **a.** 18 units

   **b.** 25 units

   **c.** 30 units

   **d.** 34 units

   **e.** 50 units

**15.** Angle *A* of rhombus *ABCD* measures 120°. If one side of the rhombus is 10 units, what is the length of the longer diagonal?

   **a.** $\sqrt{3}$ units

   **b.** 5 units

   **c.** $5\sqrt{3}$ units

   **d.** 10 units

   **e.** $10\sqrt{3}$ units

# 14 ▶ Area and Volume

## ▶ Area

The **area** of a two-dimensional figure is the amount of space within the borders of the figure. Area is always measured in square units, such as in.² or cm².

## ▶ Area Formulas for Common Shapes

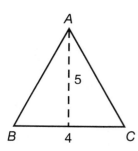

These formulas will be given to you at the SAT; you do not need to memorize them.

**Area of a Triangle:** $\frac{1}{2}bh$, where $b$ is the base of the triangle, and $h$ is its height. Triangle $ABC$ has a base of 4 units and a height of 5 units. The area of the triangle is equal to $\frac{1}{2}bh = \frac{1}{2}(4)(5) = \frac{1}{2}(20) = 10$ square units.

**Area of a Rectangle:** $lw$, where $l$ is the length of the rectangle, and $w$ is its width. Rectangle $EFGH$ has a length of 8 units and a height of 4 units. The area of the rectangle is equal to $lw = (8)(4) = 32$ square units.

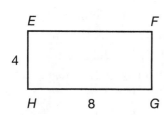

Although the area of a square can be found using the formula for the area of a rectangle, the area of a square could also be described as $s^2$, where $s$ is the length of a side of the square, since the length and width of a square are equal. If the length of one side of a square is 3 units, then the area of the square is $(3)^2 = 9$ square units.

## ▶ Volume

The **volume** of a three-dimensional figure is the amount of space the figure occupies. Volume is always measured in cubic units, such as in.$^3$ or cm$^3$.

## ▶ Volume Formulas for Common Shapes

These formulas will also be given to you at the SAT, so you do not need to memorize these either.

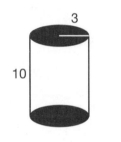

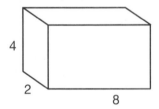

**Volume of a Cylinder:** $\pi r^2 h$, where $r$ is the radius of the cylinder, and $h$ is its height. The cylinder has a height of 10 units and a radius of 3 units. The volume of the cylinder is equal to $\pi r^2 h = \pi(3)^2(10) = \pi(9)(10) = 90\pi$ cubic units.

**Volume of a Rectangular Solid:** $lwh$, where $l$ is the length of the rectangular solid, $w$ is its width, and $h$ is its height. The rectangular solid has a length of 8 units, a width of 2 units, and a height of 4 units. The volume of the rectangular solid is $lwh = (8)(2)(4) = 64$ cubic units.

Although the volume of a cube can be found using the formula for the volume of a rectangular solid, the volume of a cube could also be described as $e^3$, where $e$ is the length of an edge of the cube. The length, width, and height of a cube are all the same, so multiplying the length, width, and height is the same as cubing any one of those measurements.

## ▶ Surface Area

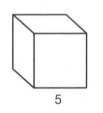

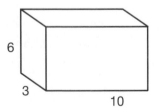

The **surface area** of a three-dimensional shape is the sum of the areas of each side of the shape. For instance, the surface area of a cube is equal to the sum of the areas of the six squares that comprise the cube. The area of one square of the cube is equal to $(5)(5) = 25$ square units. Since all six squares that comprise the cube are identical, the surface area of the square is equal to $(25)(6) = 150$ square units.

The surface area of a rectangular solid is equal to the sum of the areas of the 6 rectangles (3 pairs of congruent rectangles) that comprise the rectangular solid. The rectangular solid is composed of two rectangles that measure 6 units by 10 units, two rectangles that measure 6 units by 3 units, and two rectangles that measure 3 units by 10 units. The surface area of the rectangular solid is equal to $2(6 \times 10) + 2(6 \times 3) + 2(3 \times 10) = 2(60) + 2(18) + 2(30) = 120 + 36 + 60 = 216$ square units.

## ► Practice

**1.** If the height of a triangle is half its base, $b$, what is the area of the triangle?

   **a.** $\frac{1}{4}b$

   **b.** $\frac{1}{4}b^2$

   **c.** $\frac{1}{2}b$

   **d.** $\frac{1}{2}b^2$

   **e.** $b$

**2.** An isosceles right triangle has a hypotenuse of $x\sqrt{6}$ units. What is the area of the triangle?

   **a.** $x\sqrt{3}$ square units

   **b.** $\frac{3}{2}x^2$ square units

   **c.** $3x$ square units

   **d.** $3x^2$ square units

   **e.** $9x$ square units

**3.** Triangle $DEC$ is inscribed in rectangle $ABCD$. If side $AB = 30$ units, side $EC = 17$ units, and side $AE =$ side $EB$, what is the area of triangle $DEC$?

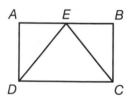

   **a.** 60 square units

   **b.** 120 square units

   **c.** 136 square units

   **d.** 240 square units

   **e.** 255 square units

**4.** What is the area of an equilateral triangle that has a perimeter of 36 units?

   **a.** $36\sqrt{3}$ square units

   **b.** 72 square units

   **c.** $72\sqrt{3}$ square units

   **d.** 108 square units

   **e.** 144 square units

**5.** If the length of $\overline{AB}$ of square $ABCD$ is $x$ units, and the length of $\overline{EF}$ is $\frac{2}{3}$ of $\overline{AD}$, what is the size of the shaded area?

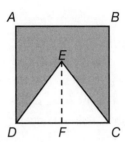

   **a.** $x^2 - \frac{1}{2}x$ square units

   **b.** $x^2 - \frac{1}{3}x$ square units

   **c.** $\frac{1}{3}x^2$ square units

   **d.** $x^2 - \frac{2}{3}x$ square units

   **e.** $\frac{2}{3}x^2$ square units

**6.** The area of a rectangle is $x^2 + 7x + 10$ square units. If the length of the rectangle is $x + 2$ units, what is the width of the rectangle?

   **a.** $x + 2$ units

   **b.** $x + 4$ units

   **c.** $x + 5$ units

   **d.** $2x + 4$ units

   **e.** $x^2 + 6x + 8$ units

**7.** If the lengths of the sides of a square are halved, the area of the new square is

   **a.** one-fourth the area of the old square.

   **b.** one-half the area of the old square.

   **c.** equal to the area of the old square.

   **d.** twice the area of the old square.

   **e.** four times the area of the old square.

**8.** The perimeter of a square is $3x - 4$ units. If the area of the square is 25 square units, what is the value of $x$?
   **a.** 5
   **b.** 6
   **c.** 7
   **d.** 8
   **e.** 20

**9.** The length of a rectangle is two less than three times its width. If the area of the rectangle is 96 cm$^2$, what is the length of the rectangle?
   **a.** 6 cm
   **b.** 8 cm
   **c.** 16 cm
   **d.** 18 cm
   **e.** 20 cm

**10.** $\overline{AC}$ is a diagonal of rectangle $ABCD$ below. If angle $ACD$ is 30° and $\overline{AC} = 20$ units, what is the area of rectangle $ABCD$?

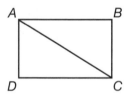

   **a.** 50 square units
   **b.** $50\sqrt{3}$ square units
   **c.** 100 square units
   **d.** $100\sqrt{3}$ square units
   **e.** 200 square units

**11.** A cylinder has a volume of $45\pi$ in.$^3$. Which of the following could be the radius and height of the cylinder?
   **a.** radius = 3 in., height = 5 in.
   **b.** radius = 3 in., height = 15 in.
   **c.** radius = 5 in., height = 3 in.
   **d.** radius = 9 in., height = 5 in.
   **e.** radius = 9 in., height = 10 in.

**12.** Terri fills with water $\frac{2}{3}$ of a glass that is 15 cm tall. If the radius of the glass is 2 cm, what volume of water is in Terri's glass?
   **a.** $10\pi$ cm$^3$
   **b.** $20\pi$ cm$^3$
   **c.** $30\pi$ cm$^3$
   **d.** $40\pi$ cm$^3$
   **e.** $60\pi$ cm$^3$

**13.** The height of cylinder B is three times the height of cylinder A, and the radius of cylinder B is $\frac{1}{3}$ the radius of cylinder A. Which of the following statements is true?
   **a.** The volume of cylinder B is $\frac{1}{9}$ the volume of cylinder A.
   **b.** The volume of cylinder B is $\frac{1}{3}$ the volume of cylinder A.
   **c.** The volume of cylinder B is the same as the volume of cylinder A.
   **d.** The volume of cylinder B is 3 times the volume of cylinder A.
   **e.** The volume of cylinder B is 9 times the volume of cylinder A.

**14.** The radius of a cylinder is $2x$ and the height of the cylinder is $8x + 2$. What is the volume of the cylinder in terms of $x$?
   **a.** $(16x^2 + 4x)\pi$
   **b.** $(16x^3 + 4x^2)\pi$
   **c.** $(32x^2 + 8x)\pi$
   **d.** $(32x^3 + 8x^2)\pi$
   **e.** $(128x^3 + 64x^2 + 8x)\pi$

**15.** The height of a cylinder is four times the radius of the cylinder. If the volume of the cylinder is $256\pi$ cm$^3$, what is the radius of the cylinder?
   **a.** 4 cm
   **b.** 8 cm
   **c.** 16 cm
   **d.** 24 cm
   **e.** 32 cm

**16.** The length of a rectangular solid is twice the sum of the width and height of the rectangular solid. If the width is equal to the height and the volume of the solid is 108 in.$^3$, what is the length of the solid?
 **a.** 3 in.
 **b.** 6 in.
 **c.** 8 in.
 **d.** 9 in.
 **e.** 12 in.

**17.** The area of one face of a cube is $9x$ square units. What is the volume of the cube?
 **a.** $27\sqrt{x}$ cubic units
 **b.** $27x$ cubic units
 **c.** $27x\sqrt{x}$ cubic units
 **d.** $27x^2$ cubic units
 **e.** $27x^3$ cubic units

**18.** The volume of rectangular solid A is equal to the volume of rectangular solid B. If the length of solid A is three times the length of solid B, and the height of solid A is twice the height of solid B, then
 **a.** the width of solid B is $\frac{1}{36}$ the width of solid A.
 **b.** the width of solid B is $\frac{1}{6}$ the width of solid A.
 **c.** the width of solid B is $\frac{1}{5}$ the width of solid A.
 **d.** the width of solid B is five times the width of solid A.
 **e.** the width of solid B is six times the width of solid A.

**19.** The volume of Stephanie's cube is equal to $64x^6$. What is the area of one face of her cube?
 **a.** $4x^2$
 **b.** $8x^2$
 **c.** $8x^3$
 **d.** $16x^4$
 **e.** $32x^3$

**20.** The length of a rectangular solid is 6 units, and the height of the solid is 12 units. If the volume of the solid is 36 cubic units, what is the width of the solid?
 **a.** $\frac{1}{2}$ units
 **b.** 2 units
 **c.** 3 units
 **d.** 6 units
 **e.** 12 units

**21.** A rectangular solid measures 4 units by 5 units by 6 units. What is the surface area of the solid?
 **a.** 60 square units
 **b.** 74 square units
 **c.** 110 square units
 **d.** 120 square units
 **e.** 148 square units

**22.** Danielle's cube has a volume of 512 in.$^3$. What is the surface area of her cube?
 **a.** 64 in.$^2$
 **b.** 132 in.$^2$
 **c.** 384 in.$^2$
 **d.** 512 in.$^2$
 **e.** 3,072 in.$^2$

**23.** The surface area of a rectangular solid is 192 cm$^2$. If the height of the solid is 4 units and the length of the solid is 12 units, what is the width of the solid?
 **a.** 2 units
 **b.** 3 units
 **c.** 4 units
 **d.** 6 units
 **e.** 12 units

**24.** The volume of a cube is $x^3$ cubic units, and the surface area of the cube is $x^3$ square units. What is the value of $x$?

   **a.** 1 unit

   **b.** 3 units

   **c.** 4 units

   **d.** 5 units

   **e.** 6 units

**25.** The width of a rectangular solid is twice the height of the solid, and the height of the solid is twice the length of the solid. If $x$ is the length of the solid, what is the surface area of the solid in terms of $x$?

   **a.** $8x^2$

   **b.** $11x^2$

   **c.** $14x^2$

   **d.** $22x^2$

   **e.** $28x^2$

# 15 ▶ Circles

**A** circle is a closed figure, but it is not a polygon. It is a 360° arc in which every point on the arc is the same distance from a single point—the center of the circle. An **arc** is a curved line that makes up part or all of a circle. If a circle is comprised of two arcs, the larger arc is called the **major arc**, and the smaller arc is called the **minor arc**.

## ▶ Circumference

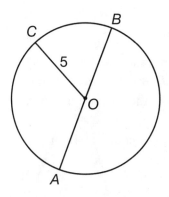

The **circumference** of a circle is the distance around the circle. The circumference of a circle is equal to $2\pi r$, where $r$ is the radius of the circle. The **radius** of a circle is the distance from the center of the circle to any point on the circle. $\overline{CO}$ is a radius of the circle. The **diameter** of a circle is the length of a straight line from a point on one side of the circle, through the center of the circle, to another point on the other side of the circle. $\overline{AB}$ is a diameter of the circle. The diameter of a circle is twice the radius. The circumference of a circle can also be given as $\pi d$, where $d$ is the diameter of the circle. The formula for circumference will be given on the SAT; you do not need to memorize it.

The circle has a radius of 5 units. Therefore, its diameter is $(2)(5) = 10$ units, and its circumference is $(2\pi)(5) = 10\pi$ units.

## ▶ Area

The **area** of a circle is the amount of space within the border of the circle. As with polygons, area is always measured in square units, such as in.$^2$ or cm$^2$. The area of a circle is equal to $\pi r^2$. This formula will also be given to you at the SAT; you do not need to memorize it.

The circle on the previous page has a radius of 5 units. Therefore, its area is equal to $\pi(5)^2 = 25\pi$ square units.

## ▶ Central Angles

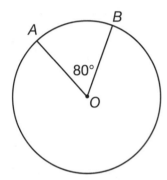

A **central angle** is an angle whose vertex is at the center of the circle and whose vertex is formed by two radii. The arc formed by the two radii is the intercepted arc of the central angle. A central angle and its intercepted arc are equal in measure.

In the circle, radii *AO* and *BO* form central angle *AOB*, which measures 80°. Therefore, intercepted (minor) arc *AB* also measures 80°. Major arc *AB*, the longer distance from *A* to *B*, measures $360 - 80 = 280°$, since there are 360° in a circle.

## ▶ Area of a Sector

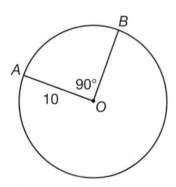

The **area of a sector** is the area of a fraction of a circle. Multiply the area of a circle by the fraction of the circle represented by the sector. The size of the sector is equal to the angle of the sector divided by 360, since there are 360° in a circle.

The circle has a sector whose angle measures 90°, and the radius of the circle is 10 units. The area of the circle is $\pi(10)^2 = 100\pi$ square units. The area of the sector is equal to a fraction of that: $\frac{90}{360}(100\pi) = \frac{1}{4}(100\pi) = 25\pi$ square units.

## ▶ Arc Length

The **length of an arc** is the distance between two points on a circle. You saw how the area of a sector is equal to a fraction of the area of a circle. The length of an arc is equal to a fraction of the circumference of a circle. Two radii form a central angle and its intercepted arc. The length of that arc is equal to the size of the central angle divided by 360, multiplied by the circumference of the circle.

The circle above has a radius of 10 units. Therefore, its circumference is $2\pi(10) = 20\pi$ units. Minor arc *AB* is formed by two radii that meet at a 90° angle. The length of arc *AB* is equal to $\frac{90}{360}(20\pi) = \frac{1}{4}(20\pi) = 5\pi$ units.

## ▶ Practice

Use the diagram below to answer questions 1–5. The diagram is not to scale.

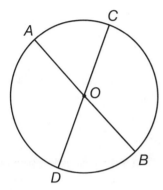

**1.** If the length of $\overline{AO}$ is 15 units, what is the circumference of the circle?

  **a.** 15π units

  **b.** 30 units

  **c.** 30π units

  **d.** 225 units

  **e.** 225π units

**2.** If the measure of angle *AOC* is 60°, what is the measure of arc *DB*?

  **a.** 30°

  **b.** 60°

  **c.** 120°

  **d.** 300°

  **e.** cannot be determined

**3.** If the length of $\overline{OD}$ is 6 units and the measure of angle *COB* is 100°, what is the length of arc *CB*?

  **a.** $\frac{10}{3}\pi$ units

  **b.** 12π units

  **c.** $\frac{50}{3}\pi$ units

  **d.** 600π units

  **e.** 1,200π units

**4.** If the area of the circle is 196π square units, what is the length of $\overline{CD}$?

  **a.** $\frac{49}{90}$ units

  **b.** 7 units

  **c.** 14 units

  **d.** 28 units

  **e.** 98 units

**5.** If the length of $\overline{OA}$ is 8 units and the measure of angle *AOC* is 50°, what is the area of sector *AOC*?

  **a.** $\frac{5}{18}\pi$ square units

  **b.** $\frac{10}{9}\pi$ square units

  **c.** $\frac{32}{25}\pi$ square units

  **d.** $\frac{80}{9}\pi$ square units

  **e.** 9π square units

Use the diagram below to answer questions 6–9. The diagram is not to scale.

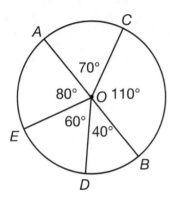

**6.** If the radius of the circle above is 12 units, which sector has an area of $24\pi$ square units?

    **a.** sector *EOD*

    **b.** sector *DOB*

    **c.** sector *BOC*

    **d.** sector *AOC*

    **e.** sector *EOA*

**7.** If the radius of the circle is 15 units, what is the area of sector *DOB*?

    **a.** $\frac{30}{9}\pi$ square units

    **b.** $22.5\pi$ square units

    **c.** $25\pi$ square units

    **d.** $45\pi$ square units

    **e.** $225\pi$ square units

**8.** If the radius of the circle is 27 units, what is the length of arc *AE*?

    **a.** $6\pi$ units

    **b.** $12\pi$ units

    **c.** $27\pi$ units

    **d.** $36\pi$ units

    **e.** $54\pi$ units

**9.** If the radius of the circle is 9 units, what is the length of arc *DB*?

    **a.** $\frac{1}{9}\pi$ units

    **b.** $1\pi$ unit

    **c.** $2\pi$ units

    **d.** $36\pi$ units

    **e.** $54\pi$ units

**10.** Jasmin draws a circle with a radius of $9x^2$. What is the area of her circle?

    **a.** $(4.5x^2)\pi$ square units

    **b.** $(3x)\pi$ square units

    **c.** $(18x^2)\pi$ square units

    **d.** $(27x^2)\pi$ square units

    **e.** $(81x^4)\pi$ square units

**11.** If the circumference of a circle triples, the area of the circle becomes

    **a.** 9 times smaller.

    **b.** 3 times smaller.

    **c.** 3 times bigger.

    **d.** 6 times bigger.

    **e.** 9 times bigger.

**12.** If the area of a circle is $(121x)\pi$ square units, what is the circumference of the circle?

    **a.** $(11\sqrt{x})\pi$ units

    **b.** $(22\sqrt{x})\pi$ units

    **c.** $(11x)\pi$ units

    **d.** $(22x)\pi$ units

    **e.** $(\frac{121}{2}x)\pi$ units

**13.** If the diameter of a circle is $8x + 6$, what is the area of the circle?

    **a.** $(4x + 3)\pi$ square units

    **b.** $(16x + 12)\pi$ square units

    **c.** $(16x^2 + 9)\pi$ square units

    **d.** $(16x^2 + 24x + 9)\pi$ square units

    **e.** $(64x^2 + 96x + 36)\pi$ square units

**14.** If the diameter of a circle is doubled, the circumference of the new circle is

    **a.** one-fourth of the circumference of the original circle.

    **b.** one-half of the circumference of the original circle.

    **c.** the same as the circumference of the original circle.

    **d.** two times the circumference of the original circle.

    **e.** four times the circumference of the original circle.

**15.** The radius of Carly's circle is $2x - 7$, and the area of her circle is $(16x + 9)\pi$. Which of the following could be the value of $x$?

    **a.** 3

    **b.** 6

    **c.** 9

    **d.** 10

    **e.** 13

**16.** If the area of a circle is $(4x^2 + 20x + 25)\pi$, what is the diameter of the circle?

    **a.** $2x + 5$

    **b.** $4x + 10$

    **c.** $(2x + 5)\pi$

    **d.** $(4x + 5)\pi$

    **e.** $2x^2 + 10x + 12.5$

**17.** If the central angle of a sector is $x°$ and the radius of the circle is $x$ units, then the area of the sector is equal to

    **a.** $\frac{x}{360}\pi$ square units

    **b.** $x\pi$ square units

    **c.** $(x^2 - \frac{x}{360})\pi$ square units

    **d.** $(\frac{x^2}{360})\pi$ square units

    **e.** $(\frac{x^3}{360})\pi$ square units

**18.** The measure of a central angle is 18°. If the length of its intercepted arc is $x\pi$ units, what is the circumference of the circle?

    **a.** $\frac{x}{20}\pi$ units

    **b.** $\frac{x}{18}\pi$ units

    **c.** $\frac{x}{10}\pi$ units

    **d.** $18x\pi$ square units

    **e.** $20x\pi$ square units

**19.** If the measure of angle $COB$ below is $3x$, what is the measure of arc $CA$?

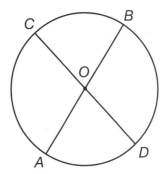

    **a.** $3x$

    **b.** $6x$

    **c.** $6x\pi$

    **d.** $12x$

    **e.** $180 - 3x$

**20.** If a circle has an area of $12\pi$ cm² and a diameter $\overline{AB}$, what is the length of arc $AB$?

    **a.** $\sqrt{3}\pi$ cm

    **b.** $3\pi$ cm

    **c.** $2\sqrt{3}\pi$ cm

    **d.** $6\pi$ cm

    **e.** $4\sqrt{3}\pi$ cm

**21.** The circumference of a circle is $16\pi$ cm. What is the area of a sector whose central angle measures 120°?

  **a.** $\frac{8}{3}\pi$ cm$^2$

  **b.** $\frac{16}{3}\pi$ cm$^2$

  **c.** $\frac{32}{3}\pi$ cm$^2$

  **d.** $\frac{64}{3}\pi$ cm$^2$

  **e.** $\frac{256}{3}\pi$ cm$^2$

Use the diagram below to answer questions 22–23. The diagram is not to scale.

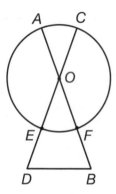

**22.** If angles $D$ and $B$ both equal 70°, what is the measure of arc $AC$?

  **a.** 35°

  **b.** 40°

  **c.** 50°

  **d.** 70°

  **e.** 110°

**23.** If $\overline{OF} = \overline{FB}$, angle $D$ = angle $B$, the radius of the circle is $6x$ and arc $EF$ is 60°, what is the perimeter of triangle $DOB$?

  **a.** $12x$

  **b.** $18x$

  **c.** $36x$

  **d.** $54x$

  **e.** cannot be determined

Use the diagram below to answer questions 24–25. The diagram is not to scale.

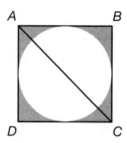

**24.** If the length of $\overline{BD}$ in square $ABCD$ is $x$ ft., what is the size of the shaded area?

  **a.** $(\frac{x^2}{2})\pi$ ft.$^2$

  **b.** $(\frac{3x^2}{4})\pi$ ft.$^2$

  **c.** $x^2 - \frac{1}{4}x^2\pi$ ft.$^2$

  **d.** $x^2 - \pi$ ft.$^2$

  **e.** $x^2 - \frac{\pi}{4}$ ft.$^2$

**25.** If the area of the circle is $25\pi$ cm$^2$, what is the length of diagonal $AD$?

  **a.** $5\sqrt{2}$ cm

  **b.** $5\sqrt{2\pi}$ cm

  **c.** $10\sqrt{2}$ cm

  **d.** $10\pi$ cm

  **e.** $10\sqrt{2\pi}$ cm

Use the diagram below to answer questions 26–27. *ABCD* is a square. The diagram is not to scale.

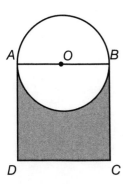

**26.** If the area of the circle is $8x^2\pi$, what is the size of the shaded area?
  **a.** $16x^2\sqrt{2} - 4x^2\pi$
  **b.** $32x^2 - 4x^2\pi$
  **c.** $16x^2\sqrt{2} - 4x\pi$
  **d.** $16x^2\sqrt{2} + 4x^2\pi$
  **e.** $32x^2 + 4x\pi$

**27.** If the area of the square is 144 square units, what is the total area of the figure?
  **a.** $144 - 72\pi$ square units
  **b.** $144 - 36\pi$ square units
  **c.** $144 - 18\pi$ square units
  **d.** $144 + 18\pi$ square units
  **e.** $144 + 36\pi$ square units

Use the diagram below to answer questions 28–30. The semicircles to the left and right of the center circle are each exactly half the size of the center circle, and the three figures are adjacent within rectangle *ABCD*. The diagram is not to scale.

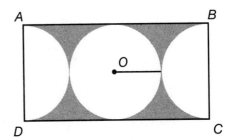

**28.** If the length of $\overline{AB}$ is $x$ units, what is the area of the center circle?
  **a.** $\frac{(x^2\pi)}{16}$ square units
  **b.** $\frac{(x^2\pi)}{4\pi}$ square units
  **c.** $x^2\pi$ square units
  **d.** $2x^2\pi$ square units
  **e.** $4x^2\pi$ square units

**29.** If the area of one semicircle is $4.5\pi$ square units, what is the area of the rectangle?
  **a.** 72 square units
  **b.** 108 square units
  **c.** 144 square units
  **d.** 162 square units
  **e.** 324 square units

**30.** If the radius of the circle is 4 units, what is the size of the shaded area?
  **a.** $72 - 16\pi$ square units
  **b.** $128 - 32\pi$ square units
  **c.** $128 - 24\pi$ square units
  **d.** $128 - 16\pi$ square units
  **e.** $112\pi$ square units

# 16 ▶ Coordinate Geometry

**T**he **coordinate plane** is the grid of boxes on which the *x*- and *y*-axes are placed and coordinate points called ordered pairs can be plotted. The points and figures that can be plotted on the plane and the operations that can be performed on them fall under the heading **coordinate geometry**.

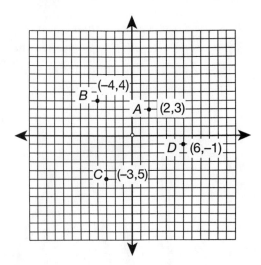

The graph is an example of the coordinate plane. The **x-axis** and the **y-axis** meet at the origin, a point with the coordinates (0,0). The **origin** is 0 units from the *x*-axis and 0 units from the *y*-axis.

Look at the point labeled *A*, with the coordinates (2,3). The *x*-coordinate is listed first in the coordinate pair. The *x* value of a point is the distance from the *y*-axis to that point. Point *A* is 2 units from the *y*-axis, so its *x* value is 2. The *y*-coordinate is listed second in the coordinate pair. The *y* value of a point is the distance from the *x*-axis to that point. Point *A* is 3 units from the *x*-axis, so its *y* value is 3.

What are the coordinates of point *B*? Point *B* is –4 units from the *y*-axis and 4 units from the *x*-axis. The coordinates of point *B* are (–4,4).

The coordinate plane is divided into 4 sections, or quadrants. The points in quadrant I, the top right corner of the plane, have positive values for both *x* and *y*. Point *A* is in quadrant I and its *x* and *y* values are both positive. The points in quadrant II, the top left corner of the plane, have negative values for *x* and positive values for *y*. Point *B* is in quadrant II and its *x* value is negative, while its *y* value is positive. The points in quadrant III, the bottom left corner of the plane, have negative values for both *x* and *y*, and the points in quadrant IV, the bottom right corner of the plane, have positive values for *x* and negative values for *y*.

## ▶ Slope

When two points on the coordinate plane are connected, a line is formed. The **slope** of a line is the difference between the *y* values of two points divided by the difference between the *x* values of those two points. When the equation of a line is written in the form $y = mx + b$, the value of *m* is the slope of the line.

If both the *y* value and the *x* value increase from one point to another, or, if both the *y* value and the *x* value decrease from one point to another, the slope of the line is positive. If the *y* value increases and the *x* value decreases from one point to another, or, if the *y* value decreases and the *x* value increases from one point to another, the slope of the line is negative.

A horizontal line has a slope of 0. Lines such as $y = 3$, $y = -2$, or $y = c$, where *c* is any constant, are lines with slopes of 0.

A vertical line has no slope. Lines such as $x = 3$, $x = -2$, or $x = c$ are lines with no slopes.

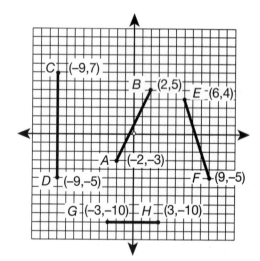

The slope of the line *AB* is equal to $\frac{(5-(-3))}{(2-(-2))} = \frac{8}{4} = 2$. The slope of line *AB* is 2. The slope of line *EF* is equal to $\frac{((-5)-4)}{(9-6)} = \frac{-9}{3} = -3$. Line *GH* is a horizontal line; there is no change in the *y* values from point *G* to point *H*. This line has a slope of 0. Line *CD* is a vertical line; there is no change in the *x* values from point *C* to point *D*. This line has no slope.

Parallel lines have the same slope. Perpendicular lines have slopes that are negative reciprocals of each other. Lines given by the equations $y = 3x + 5$ and $y = 3x - 2$ are parallel, while the line given by the equation $y = -\frac{1}{3}x + 1$ is perpendicular to those lines.

## ▶ Midpoint

The **midpoint** of a line segment is the coordinates of the point that falls exactly in the middle of the line segment. If $(a,c)$ is one endpoint of a line segment and $(b,d)$ is the other endpoint, the midpoint of the line segment is equal to $(\frac{a+b}{2}, \frac{c+d}{2})$. In other words, the midpoint of a line segment is equal to the average of the $x$ values of the endpoints and the average of the $y$ values of the endpoints.

What is the midpoint of a line segment with endpoints at $(1,5)$ and $(-3,3)$?

Using the midpoint formula, the midpoint of this line is equal to $(\frac{1+(-3)}{2}, \frac{5+3}{2}) = (\frac{-2}{2}, \frac{8}{2}) = (-1,4)$.

## ▶ Distance

To find the **distance** between two points, use the formula below. The variable $x_1$ represents the $x$-coordinate of the first point, $x_2$ represents the $x$-coordinate of the second point, $y_1$ represents the $y$-coordinate of the first point, and $y_2$ represents the $y$-coordinate of the second point:

$$D = \sqrt{((x_2 - x_1)^2 + (y_2 - y_1)^2)}$$

What is the distance between the points $(-2,8)$ and $(4,-2)$?

Substitute these values into the formula:

$$D = \sqrt{((4 - (-2))^2 + ((-2) - 8)^2)}$$

$D = \sqrt{((4 + 2)^2 + (-2 - 8)^2)}$
$D = \sqrt{((6)^2 + (-10)^2)}$
$D = \sqrt{(36 + 100)}$
$D = \sqrt{136}$
$D = 2\sqrt{34}$

## ▶ Practice

**1.** The endpoints of a line segment are $(-3,6)$ and $(7,4)$. What is the slope of this line?
- **a.** −5
- **b.** $-\frac{1}{5}$
- **c.** $\frac{1}{5}$
- **d.** 5
- **e.** 10

**2.** The endpoints of a line segment are $(5,-5)$ and $(-5,-5)$. What is the slope of this line?
- **a.** −10
- **b.** −5
- **c.** 0
- **d.** 5
- **e.** This line has no slope.

**3.** What is the slope of a line segment with end-points at (−1,2) and (1,10)?

   **a.** −4

   **b.** −$\frac{1}{4}$

   **c.** $\frac{1}{4}$

   **d.** 4

   **e.** This line has no slope.

Use the diagram below to answer questions 4–5.

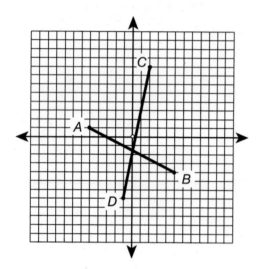

**4.** What is the slope of line segment *AB*?

   **a.** −$\frac{1}{2}$

   **b.** −1

   **c.** −2

   **d.** −5

   **e.** −10

**5.** What is the slope of line segment *CD*?

   **a.** $\frac{1}{5}$

   **b.** 1

   **c.** 3

   **d.** 5

   **e.** 15

**6.** What is the midpoint of a line segment with end-points at (0,−8) and (−8,0)?

   **a.** (−8,−8)

   **b.** (−4,−4)

   **c.** (−1,−1)

   **d.** (4,4)

   **e.** (8,8)

**7.** What is the midpoint of a line segment with end-points at (6,−4) and (15,8)?

   **a.** (9,4)

   **b.** (9,12)

   **c.** (10.5,2)

   **d.** (12,2)

   **e.** (12,9)

**8.** The endpoints of a line segment are (0,−4) and (0,4). What is the midpoint of this line?

   **a.** (0,0)

   **b.** (0,−2)

   **c.** (0,2)

   **d.** (−4,4)

   **e.** This line has no midpoint.

Use the diagram below to answer questions 9–10.

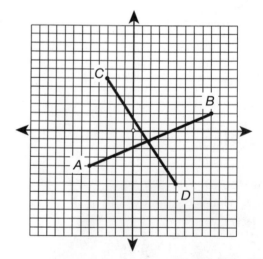

**9.** What is the midpoint of line segment *AB*?

   **a.** (1.5,–1)

   **b.** (0,2)

   **c.** (4,0)

   **d.** (2,–1)

   **e.** (4,–2)

**10.** What is the midpoint of line segment *CD*?

   **a.** (1,0)

   **b.** (1,–3)

   **c.** (2,0)

   **d.** (2,–1)

   **e.** (4,6)

**11.** What is the distance from the point (–6,2) to the point (2,17)?

   **a.** $3\sqrt{41}$ units

   **b.** $\sqrt{229}$ units

   **c.** 17 units

   **d.** $\sqrt{365}$ units

   **e.** $5\sqrt{17}$ units

**12.** What is the distance from the point (0,–4) to the point (4,4)?

   **a.** $5\sqrt{2}$ units

   **b.** 4 units

   **c.** $4\sqrt{2}$ units

   **d.** $4\sqrt{3}$ units

   **e.** $4\sqrt{5}$ units

**13.** What is the distance from the point (3,8) to the point (7,–6)?

   **a.** $2\sqrt{5}$ units

   **b.** $2\sqrt{47}$ units

   **c.** $2\sqrt{51}$ units

   **d.** $2\sqrt{53}$ units

   **e.** $2\sqrt{74}$ units

Use the diagram below to answer questions 14–15.

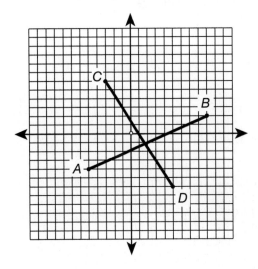

**14.** What is the distance from the point *A* to point *B*?

   **a.** $2\sqrt{13}$ units

   **b.** 10 units

   **c.** $10\sqrt{2}$ units

   **d.** $2\sqrt{58}$ units

   **e.** 20 units

**15.** What is the distance from the point *C* to point *D*?

   **a.** 8 units

   **b.** 10 units

   **c.** $2\sqrt{37}$ units

   **d.** $4\sqrt{13}$ units

   **e.** $16\sqrt{3}$ units

**16.** What is the slope of the line given by the equation $5y = -3x + 6$?

   **a.** –3

   **b.** $-\frac{5}{3}$

   **c.** $-\frac{3}{5}$

   **d.** $\frac{6}{5}$

   **e.** 5

**17.** Which of the following lines is parallel to the line given by the equation $y = -2x + 4$?

  **a.** $y = -2x - 4$

  **b.** $y = -\frac{1}{2}x + 4$

  **c.** $y = -\frac{1}{2}x - 4$

  **d.** $y = \frac{1}{2}x + 4$

  **e.** $y = 2x - 4$

**18.** Which of the following lines is perpendicular to the line given by the equation $y = -\frac{1}{6}x + 8$?

  **a.** $y = -6x - 8$

  **b.** $y = -6x + 8$

  **c.** $y = -\frac{1}{6}x - 8$

  **d.** $y = \frac{1}{6}x - 8$

  **e.** $y = 6x + 8$

**19.** Which of the following lines is parallel to the line given by the equation $4y = 6x - 6$?

  **a.** $y = -\frac{3}{2}x + 6$

  **b.** $y = -\frac{2}{3}x + 6$

  **c.** $y = \frac{2}{3}x + 6$

  **d.** $y = \frac{3}{2}x + 3$

  **e.** $y = 2x - 10$

**20.** Which of the following lines is perpendicular to the line given by the equation $-2y = -8x + 10$?

  **a.** $y = -4x - 5$

  **b.** $y = -\frac{1}{4}x + 5$

  **c.** $y = \frac{1}{4}x - 5$

  **d.** $y = \frac{1}{8}x + 5$

  **e.** $y = 4x - 5$

**21.** What is the distance from the point $(-x,y)$ to the point $(x,-y)$?

  **a.** $(x + y)$ units

  **b.** $\sqrt{(x + y)}$ units

  **c.** $(x^2 + y^2)$ units

  **d.** $\sqrt{(x^2 + y^2)}$ units

  **e.** $2\sqrt{(x^2 + y^2)}$ units

**22.** Two perpendicular lines intersect at the point $(1,5)$. If the slope of one line is 3, what is the equation of the other line?

  **a.** $y = -3x + 8$

  **b.** $y = -\frac{1}{3}x + 2$

  **c.** $y = -\frac{1}{3}x + \frac{16}{3}$

  **d.** $y = \frac{1}{3}x + \frac{14}{3}$

  **e.** $y = 3x + 2$

**23.** What is the midpoint of a line with endpoints at $(2x + 3, y - 4)$ and $(10x - 1, 3y + 6)$?

  **a.** $(x + 1, y + 1)$

  **b.** $(\frac{3}{2}x + \frac{3}{2}, \frac{5}{2}y - \frac{5}{2})$

  **c.** $(6x + 1, 2y + 1)$

  **d.** $(8x - 4, 2y + 10)$

  **e.** $(12x + 2, 4y + 2)$

**24.** Which of the following is the product of the slopes of perpendicular lines?

  **a.** $-1$

  **b.** $-\frac{1}{2}$

  **c.** $0$

  **d.** $\frac{1}{2}$

  **e.** $1$

**25.** Line $A$ is perpendicular to line $B$. If the slope of line $A$ is multiplied by 4, what must the slope of line $B$ be multiplied by in order for the lines to still be perpendicular?

  **a.** $-4$

  **b.** $-\frac{1}{4}$

  **c.** $-\frac{1}{16}$

  **d.** $\frac{1}{4}$

  **e.** $4$

# Posttest ▶

f you have completed all lessons in this book, then you are ready to take the posttest to measure your progress. The posttest has 25 multiple-choice questions covering the topics you studied in this book, of which 15 are the multiple-choice questions and 10 are the grid-ins. While the format of the posttest is similar to that of the pretest, the questions are different.

Take as much time as you need to complete the posttest. When you are finished, check your answers with the answer key at the end of the book. Once you know your score on the posttest, compare the results with the pretest. If you scored better on the posttest than you did on the pretest, congratulations! You have profited from your hard work. At this point, you should look at the questions you missed, if any. Do you know why you missed the question, or do you need to go back to the lesson and review the concept?

If your score on the posttest doesn't show much improvement, take a second look at the questions you missed. Did you miss a question because of an error you made? If you can figure out why you missed the problem, then you understand the concept and just need to concentrate more on accuracy when taking a test. If you missed a question because you did not know how to work the problem, go back to the lessons and spend more time working that type of problem. You need a solid foundation in algebra and geometry if you plan to do well on the math section of the SAT!

# POSTTEST

1. ⓐ ⓑ ⓒ ⓓ ⓔ
2. ⓐ ⓑ ⓒ ⓓ ⓔ
3. ⓐ ⓑ ⓒ ⓓ ⓔ
4. ⓐ ⓑ ⓒ ⓓ ⓔ
5. ⓐ ⓑ ⓒ ⓓ ⓔ

6. ⓐ ⓑ ⓒ ⓓ ⓔ
7. ⓐ ⓑ ⓒ ⓓ ⓔ
8. ⓐ ⓑ ⓒ ⓓ ⓔ
9. ⓐ ⓑ ⓒ ⓓ ⓔ
10. ⓐ ⓑ ⓒ ⓓ ⓔ

11. ⓐ ⓑ ⓒ ⓓ ⓔ
12. ⓐ ⓑ ⓒ ⓓ ⓔ
13. ⓐ ⓑ ⓒ ⓓ ⓔ
14. ⓐ ⓑ ⓒ ⓓ ⓔ
15. ⓐ ⓑ ⓒ ⓓ ⓔ

16.

17.

18.

19.

20.

21.

22.

23.

24.

25.

**1.** The expression $\frac{(x^2 + 4x - 12)}{(x^2 - 8x + 12)}$ is equivalent to

  **a.** $\frac{1}{2}$

  **b.** $x - 2$

  **c.** $-\frac{1}{4}x$

  **d.** $\frac{x + 2}{x - 2}$

  **e.** $\frac{x + 6}{x - 6}$

**2.** In the diagram below, if the radius of the circle is 25 units, what is the length of arc *AB*?

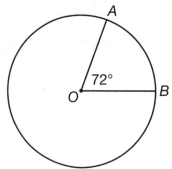

  **a.** $\frac{2}{5}\pi$

  **b.** $5\pi$

  **c.** $10\pi$

  **d.** $50\pi$

  **e.** $250\pi$

**3.** Which of the following statements is true of the graph below?

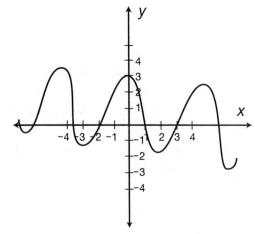

  **a.** The graphed equation is not a function.

  **b.** There are at least eight different values for which $f(x) = 1$.

  **c.** There are no values greater than 4 in the domain of the function.

  **d.** The range of the function contains no values between −2 and 1.

  **e.** There are eight *y*-intercepts for the equation.

**4.** What is the midpoint of a line segment with endpoints at (−1,4) and (13,12)?

  **a.** (5.5,4.5)

  **b.** (6,8)

  **c.** (7,4)

  **d.** (7,8)

  **e.** (12,16)

**5.** Given $\frac{x}{yz} = a$, if $x$ is doubled, then for the value of $a$ to remain the same,
   a. the value of $y$ must be halved and the value of $z$ must be halved.
   b. either the value of $y$ must be halved, or, the value of $z$ must be halved.
   c. the value of $y$ must be doubled and the value of $z$ must be doubled.
   d. either the value of $y$ must be doubled or the value of $z$ must be doubled.
   e. the value of $y$ must be doubled and the value of $z$ must be halved.

**6.** Which of the following is equivalent to $a^1a^{-2}a^3b^{-1}b^2b^{-3}$?
   a. $1$
   b. $ab$
   c. $a^2b^2$
   d. $\frac{a}{b}$
   e. $\frac{(a^2)}{(b^2)}$

**7.** If $a + \frac{5}{3} = \frac{4a}{a-2}$, what are the values of $a$?
   a. $a = -1, a = 10$
   b. $a = 1, a = -10$
   c. $a = 2, a = -5$
   d. $a = -2, a = 5$
   e. $a = -5, a = 0$

**8.** Which of the following points is in the solution set of $4y + 6 > 3x + 15$?
   a. $(0, -4)$
   b. $(3, 4)$
   c. $(4, 3)$
   d. $(4, 6)$
   e. $(5, 6)$

**9.** If the circle inscribed in square $ABCD$ has a radius of $r$, what is the size of the shaded area in terms of $r$?

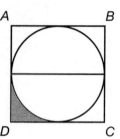

   a. $r^2 - \pi r^2$
   b. $2r - \pi r^2$
   c. $\frac{r}{2} - \frac{\pi r^2}{4}$
   d. $r^2 - \frac{\pi r^2}{4}$
   e. $\frac{(r^2 - \pi r^2)}{4}$

**10.** Based on the diagram, if lines $L$ and $M$ are parallel, which of the following equations is NOT necessarily true?

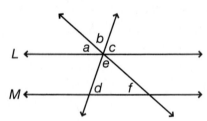

   a. $a + b + c = d + e + f$
   b. $a + c = 180 - e$
   c. $a + e + c = 180$
   d. $b + c = e + f$
   e. $a + b + c + d + e + f = 360$

**11.** Which of the following lines is perpendicular to the line $4y + 3x = 12$?
   **a.** $y = \frac{1}{3}x + 12$
   **b.** $y = -\frac{3}{4}x + 3$
   **c.** $y = \frac{3}{4}x - 3$
   **d.** $y = -\frac{4}{3}x + 4$
   **e.** $y = \frac{4}{3}x - 4$

**12.** A circle has a circumference equal to $g$. If the area of the circle is tripled, what is the new area of the circle in terms of $g$?
   **a.** $3\pi g^2$
   **b.** $\frac{g^2}{4}\pi$
   **c.** $\frac{3g^2}{4}\pi$
   **d.** $\frac{9g^2}{4}\pi$
   **e.** $\frac{9g^2}{4}\pi^2$

**13.** Compared to the graph of $y = x^2$, the graph of $y = (x - 2)^2 - 2$ is
   **a.** shifted 2 units right and 2 units down.
   **b.** shifted 2 units left and 2 units down.
   **c.** shifted 2 units right and 2 units up.
   **d.** shifted 2 units left and 2 units up.
   **e.** shifted 4 units left and 2 units down.

**14.** Aiden sketches a right triangle and labels it $ABC$. Angle $B$ is 90° and the tangent of angle $A$ is 1. If the length of side $AB$ is 10 units, what is the length of side $AC$?
   **a.** 5 units
   **b.** 10 units
   **c.** $10\sqrt{2}$ units
   **d.** $10\sqrt{3}$ units
   **e.** 20 units

**15.** What is the quotient of $\frac{(12x^4 + 21x^3 - 3x^2)}{3x^2}$?
   **a.** $4x^2 + 7x$
   **b.** $4x^2 + 7x - 1$
   **c.** $9x^2 + 18x$
   **d.** $9x^2 + 18x - 1$
   **e.** $36x^6 + 62x^5 - 9x^4$

**16.** If the area of a sector of a circle is $8\pi$ and the angle formed by the two radii of the sector is equal to 80°, what is the length of the radius of the circle?

**17.** Triangles $ABC$ and $DEF$ are similar. Each side of $ABC$ is three times the length of its corresponding side of triangle $DEF$. If the area of triangle $ABC$ is 72 square units, what is the area of triangle $DEF$ in square units?

**18.** If $a^{\frac{3}{2}} = 512$, then what is $a^{\frac{2}{3}}$?

**19.** Bria and Lindsay play tennis on a rectangular court with an area of 2,000 ft.$^2$. If the length of the court is 80 ft., what is the perimeter of the court in feet?

**20.** Marie is filling a box with books. The box has a length of 16 in., a width of 10 in., and a height of 8 in. If every book has a width of 5 in., a length of 8 in., and a height of 1 in., how many books can fit in the box?

**21.** When $x = -4$, what is the value of $\frac{x^2 + 3x}{8 - 4}$?

**22.** In the diagram below, line *CD* is a tangent and line *EO* is a secant. If arc *AB* = 60° and the radius of the circle is 7 units, what is the length of secant *EO*?

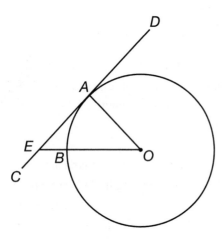

**23.** If $-3a - 9b = -6$ and $5a + 6b = -8$, what is the value of *b*?

**24.** Each term in the sequence below is twice the previous term. What is the seventh term of the sequence below?

6, 12, 24, 48, . . .

**25.** Find the positive value of *x* that makes the expression $\frac{(x^2 + 9x)}{(x^2 + 6x - 27)}$ undefined.

# Answers ▶

## ▶ Pretest

**1. d.** The expression is undefined when the denominator of the expression is equal to 0. Factor $(6x^2 + 20x + 6)$ into $(3x + 1)(2x + 6)$ and set each factor equal to 0; $3x + 1 = 0$, $3x = -1$, $x = -\frac{1}{3}$; $2x + 6 = 0$, $2x = -6$, $x = -3$.

**2. c.** Each term in the sequence is five times the term before it. The first term in the pattern is 4, or $4 \times 5^0$. The second term in the pattern is 20, or $4 \times 5^1$. Every term in the pattern is 4 times a power of 5. The exponent of 5 is equal to one less than the position of the term in the pattern. Therefore, the exponent of 5 for the eighth term is one less than 8: $4 \times 5^7$.

**3. c.** The area of a circle is equal to $\pi r^2$, where $r$ is the radius of the circle. Therefore, the radius of this circle is equal to $\sqrt{64} = 8$ ft. The circumference of a circle is equal to $2\pi r$; $2\pi(8) = 16\pi$ ft.

**4. b.** The equation $y = 2$ is the equation of horizontal line that crosses the $y$-axis at $(0,2)$. Horizontal lines have a slope of 0. This line is a function, since it passes the vertical line test: When graphed, a vertical line can be drawn through the graph of $y = 2$ at any point and that vertical line

will cross the graphed line in only one place. The domain of the function is infinite, but all $x$ values yield the same $y$ value: 2. Therefore, the range of $y = 2$ is 2.

**5. d.** If the numerator of a fraction contains a term with a negative exponent, move that term to the denominator. In the same way, if the denominator of a fraction contains a term with a negative exponent, move that term to the numerator. Therefore, the given expression is equal to $(2^{\frac{a}{b}})(2^{\frac{b}{a}})$. The $a$ and $b$ terms cancel, leaving $(2)(2) = 4$.

**6. c.** $-4(x - 1) = -4x + 4$ and $2(x + 1) = 2x + 2$. Since $-4x + 4 \leq 2x + 2$, subtract $2x$ from both sides of the inequality and subtract 4 from both sides of the inequality, leaving: $-6x \leq -2$. Divide both sides by $-6$ and change the direction of the inequality sign: $x \geq \frac{1}{3}$.

**7. b.** The area of a square is equal to the length of one side squared. Therefore, the length of side $BC$ is equal to the square root of the area of rectangle $BCEF$: $\sqrt{20} = 2\sqrt{5}$. The area of a rectangle is equal to the product of its length and width. The length of rectangle $ABCD$ is 10 units and the width is $2\sqrt{5}$ units, since rectangle $ABCD$ shares side $BC$ with square $BCEF$. Therefore, the area of rectangle $ABCD =$ $(10)(2\sqrt{5}) = 20\sqrt{5}$ square units.

**8. a.** The stack of compact discs, or a stack of circles, forms a cylinder. The volume of a cylinder is equal to $\pi r^2 h$, where $r$ is the radius of the cylinder (or one of the circles that forms it) and $h$ is the height of the cylinder. The radius of a disc is half its diameter: $\frac{12}{2} = 6$ cm. Since each disc has a height of 2 mm, the height of the cylinder is $(10)(2) = 20$ mm. Convert this measure to centimeters, since the radius of the disc is given in centimeters. $\frac{(20 \text{ mm})}{10} = 2$ cm. The volume of the cylinder $= \pi(6)^2(2) = 72\pi$ cm$^3$.

**9. a.** A parallelogram is a quadrilateral with two pairs of parallel sides. All rectangles, rhombuses, and squares are parallelograms. A rectangle is a parallelogram with four right angles. All rectangles are parallelograms, but not all parallelograms are rectangles. A rhombus is a parallelogram with four equal sides. All rhombuses are parallelograms, but not all rhombuses are rectangles and not all parallelograms are rhombuses. A square is a parallelogram with four right angles and four equal sides. All squares are rectangles, rhombuses, and parallelograms, but not all rectangles are squares, not all rhombuses are squares, and not all parallelograms are squares.

**10. e.** Factor the expression $3x^2 - 3x - 18$; $3x^2$ is equal to $3x$ multiplied by $x$. Find two numbers that multiply to $-18$ and whose difference is $-3$ after one of them is multiplied by 3; $3x^2 - 3x - 18$ factors into $(3x + 6)(x - 3)$. Set each factor equal to 0 and solve for $x$; $3x + 6 = 0$, $3x = -6$, $x = -2$; $x - 3 = 0$, $x = 3$.

**11. b.** Since $ABCD$ is a parallelogram, lines $AB$ and $CD$ are parallel to each other and lines $AC$ and $BD$ are parallel to each other. Opposite angles in a parallelogram are equal; therefore, the angle labeled $16x$ and angle $CAB$ are equal. Since angle $CAB$

and the angle labeled $9x + 5$ are supplementary, the angle labeled $16x$ and the angle labeled $9x + 5$ are supplementary: $9x + 5 + 16x = 180$, $25x = 175$, $x = 7$. Therefore, the measure of the angle labeled $9x + 5$ is $9(7) + 5 = 63 + 5 = 68°$. Since that angle and angle $ABD$ are alternating angles, their measures are equal. The measure of angle $ABD$ is also $68°$.

**12. a.** The ratio of a side of $DE$ to a side of $AB$ is 4 to 10, or $\frac{4}{10}$, which equals $\frac{2}{5}$. The area of each triangle is equal to $\frac{1}{2}$(base)(height). Since the base and the height of triangle $DEF$ are $\frac{2}{5}$ the base and height of triangle $ABC$, the area of triangle $DEF$ will be $(\frac{2}{5})(\frac{2}{5}) = \frac{4}{25}$ the area of triangle $ABC$.

**13. d.** To find the turning point of a parabola, find the value that makes the $x$ term of the equation equal to 0. Then, use that value of $x$ to find the value of $y$. Three of the given equations have $x$ values that have positive numbers added to them before squaring. For each of these (choices **a**, **c**, and **d**), the value of $x$ must be negative in order for that term of the equation to be equal to 0. For example, $x$ must equal $-1$ for the $x$ term in choices **a** and **c** to equal 0. Therefore, the turning points of these parabolas will be in either the second or third quadrants of the coordinate plane. The equations in choices **a** and **c** contain negative constants. These equations will have their turning points below the $x$-axis, in the third quadrant. Only choice **d**, $y = -(x + 2)^2 + 1$, will have a turning point with a negative $x$ value and a positive $y$ value, placing the turning point in the second quadrant of the coordinate plane.

**14. d.** Rewrite the equation in terms of $b$. Multiply both sides of the equation by 4, then add 4 to both sides of the equation: $a = \frac{7b - 4}{4}$, $4a = 7b - 4$, $4a + 4 = 7b$. Finally, divide both sides by 7; $b = \frac{4a + 4}{7}$.

**15. e.** The area of a triangle is equal to half the product of its base and height. Each triangle has a base of 4 units. If a line is drawn from one vertex of the triangle to its opposite base, that line would be perpendicular to that base and bisect it, splitting the equilateral triangle into two congruent, 30-60-90 right triangles. The length of the shorter base is half the length of the hypotenuse: $\frac{1}{2}(4) = 2$. The length of the longer base is $\sqrt{3}$ times that: $2\sqrt{3}$. Since the longer base of the right triangle is the height of the equilateral triangle, the area of the equilateral triangle is $\frac{1}{2}(4)(2\sqrt{3}) = 4\sqrt{3}$. Since the area of one equilateral triangle is $4\sqrt{3}$, the area of all six, and the area of the hexagon, is $(4\sqrt{3})(6) = 24\sqrt{3}$.

**16.** **64** A cube has 6 identical faces. If the total surface area of a cube is 96 square centimeters, then the area of one face of the cube is $\frac{96}{6} = 16$ cm$^2$. Each face of a cube is a square. Since the area of a square is equal to the length of one side of the square multiplied by itself, the length of one edge of the cube is equal to $\sqrt{16} = 4$ cm. The volume of a cube is equal to $e^3$, where $e$ is the length of one edge of the cube. Therefore, the volume of the cube is $4^3 = 64$ cm$^3$.

**17.** **20** To find the distance between two points, square the difference between the $x$ values and square the difference between the $y$ values. Then, add the two differences and take the square root: $(5 - (-7))^2 = 12^2 = 144$ and $(12 - (-4))^2 = 16^2 = 256$; $144 + 256 = 400$, $\sqrt{400} = 20$.

**18.** **1,800** The sum of the interior angles of a polygon is equal to $(180)(s-2)$, where $s$ is the number of sides of the polygon. Since Steve's polygon has 12 sides, the sum of the interior angles is equal to $(180)(12-2) = (180)(10) = 1,800$.

**19.** **42** Substitute 36 for $b$: $\frac{\sqrt{36+13}}{36}36\sqrt{36}$. The sum of 36 and 13 is 49, and the 36 in the denominator cancels with the 36 in front of $\sqrt{36}$, leaving $(\sqrt{49})(\sqrt{36}) = (7)(6) = 42$.

**20.** **50** Angles $DCA$ and $ACB$ form a line; therefore, the measures of these angles add to 180°; $180 - DCA = ACB$, $180 - 115 = 65°$. Since sides $AC$ and $AB$ are congruent, triangle $ABC$ is isosceles. The angles opposite the congruent sides are congruent. Therefore, angle $ABC$ is also 65°. Since there are 180° in a triangle, the measure of angle $A = 180 - 65 - 65 = 50°$.

**21.** **0.25** Solve the given equation for $w$. Subtract 3 from both sides of the equation and then multiply both sides by $\frac{3}{2}$; $\frac{2w}{3} + 3 = 7$, $\frac{2w}{3} = 4$, $w = 6$. Substitute 6 for $w$ in the second expression; $\frac{3}{2w} = \frac{3}{2}(6) = \frac{1}{4}$. Alternatively, notice that $\frac{2w}{3} = 4$. The reciprocal of $\frac{2w}{3}, \frac{3}{2w}$, will be equal to the reciprocal of the value of $\frac{2w}{3}$. Since $\frac{2w}{3} = 4$, $\frac{3}{2w}$ is equal to $\frac{1}{4}$.

**22.** **6** Cross multiply and solve for $x$: $(2x + 8)(6) = (5)(5x - 6)$, $12x + 48 = 25x - 30$, $78 = 13x$, $x = 6$.

**23.** **38** Since sides $OC$ and $OB$ of triangle $OBC$ are congruent, the angles opposite these sides, angles $OBC$ and $OCB$, are congruent. Therefore, angle $OCB$ is also 71°. Since there are 180° in a triangle, angle $COB$ is equal to $180 - (71 + 71) = 180 - 142 = 38°$. Angles $COB$ and $AOD$ are vertical angles; therefore, angle $AOD$ is also 38°. Angle $AOD$ is a central angle of the circle (its vertex is at the center of the circle). The measure of the intercepted arc of a central angle is equal to the measure of the central angle. Since angle $AOD$ is 38°, arc $AD = 38°$.

**24.** **25** Substitute $-2$ for $a$. $((\frac{7}{5}(-2)^2) + (\frac{3}{10}(-2)))^{-2}$ $= (\frac{7}{5}(4) + -\frac{3}{20})^{-2} = (\frac{7}{20} - \frac{3}{20})^{-2} = (\frac{4}{20})^{-2} = (\frac{20}{4})^2 = 5^2 = 25$.

**25.** **32** The base opposite the 30° angle of a 30-60-90 right triangle is the shortest side of the triangle. The longer base is $\sqrt{3}$ times the length of the shorter base, and the hypotenuse is twice the length of the shorter base. In triangle $ABC$, $\overline{BC}$ is the longer base, since it is opposite the 60° angle. Since $\overline{BC}$ is $16\sqrt{3}$, $\overline{AB}$ is equal to $\frac{16\sqrt{3}}{\sqrt{3}} = 16$, and the hypotenuse of the triangle, $\overline{AC}$, is equal to $(16)(2) = 32$.

## ► Chapter 1

**1. d.** Subtract the like terms by subtracting the coefficients of the terms: $9a - 5a = 4a$. $4a$ and $12a^2$ are not like terms, so they cannot be combined any further; $9a + 12a^2 - 5a = 12a^2 + 4a$.

**2. a.** Multiply the coefficients of the terms in the numerator, and add the exponents of the bases: $(3a)(4a) = 12a^2$. Do the same with the terms in the denominator: $[6(6a^2)] = 36a^2$. Finally, divide the numerator by the denominator. Divide the coefficients of the terms and subtract the exponents of the bases: $\frac{(12a^2)}{(36a^2)} = \frac{1}{3}$.

**3. d.** The terms $5a$ and $7b$ have unlike bases; they cannot be combined any further. Add the terms in the denominator; $b + 2b = 3b$. Divide the $b$ term in the numerator by the $3b$ in the denominator; $\frac{b}{3b} = \frac{1}{3}$. $(5a + 7b)(\frac{1}{3}) = \frac{5a + 7b}{3}$.

**4. b.** Multiply $2x^2$ and $4y^2$ by multiplying the coefficients of the terms: $(2x^2)(4y^2) = 8\,x^2y^2$. $8x^2y^2$ and $6x^2y^2$ have like bases, so they can be added. Add the coefficients: $8x^2y^2 + 6x^2y^2 = 14x^2y^2$.

**5. c.** Substitute 3 for each instance of $x$ in the expression: $2(3)^2 - 5(3) + 3 = 2(9) - 15 + 3 = 18 - 15 + 3 = 6$.

**6. b.** Substitute $-2$ for each instance of $a$ in the expression: $\frac{7(-2)}{((-2)^2 + (-2))} = \frac{-14}{(4-2)} = -\frac{14}{2} = -7$.

**7. e.** Substitute $-2$ for each instance of $x$ in the expression: $\frac{(y^2)}{(-2)^2} + \frac{y}{(-2(-2))} = \frac{y^2}{4} + \frac{y}{4} = \frac{(y^2 + y)}{4}$.

**8. e.** Substitute 6 for each instance of $a$ in the expression: $\frac{4(6)((6) + r)}{6r} = \frac{24(6 + r)}{6r} = \frac{4(6 + r)}{r} = \frac{24 + 4r}{r}$. The expression cannot be simplified any further.

**9. c.** Substitute 3 for each instance of $a$ in the expression: $(4(3)^2)(3b^3 + (3)) - b^3 = (4)(9)(3b^3 + 3) - b^3 = 36(3b^3 + 3) - b^3 = 108b^3 + 108 - b^3 = 107b^3 + 108$.

**10. a.** Substitute 1 for each instance of $c$ and substitute 4 for each instance of $d$ in the expression: $\frac{((1)(4))^2}{(1) + (4)} = \frac{4^2}{5} = \frac{16}{5}$.

**11. c.** Substitute 2 for each instance of $x$ and substitute 3 for each instance of $y$ in the expression: $\frac{6(2)^2}{2(3)^2} + \frac{4(2)}{3(3)} = \frac{(6)(4)}{(2)(9)} + \frac{8}{9} = \frac{24}{18} + \frac{8}{9} = \frac{12}{9} + \frac{8}{9} = \frac{20}{9}$.

**12. c.** Substitute 1 for each instance of $a$ and substitute $-1$ for each instance of $b$ in the expression: $(1)(-1) + \frac{1}{-1} + (1)^2 - (-1)^2 = -1 + (-1) + 1 - 1 = -2$.

**13. c.** Isolate $g$ on one side of the equation. Multiply both sides of the equation by $\frac{2}{3}$: $(\frac{2}{3})(\frac{3}{2g}) = (9h - 15)\frac{2}{3}$, $g = 6h - 10$.

**14. a.** Isolate $a$ on one side of the equation. Subtract $20b$ from both sides of the equation and divide by 7: $7a + 20b = 28 - b$, $7a = 28 - 21b$, $a = 4 - 3b$.

**15. d.** Isolate $y$ on one side of the equation. Multiply the $(\frac{x}{y} + 1)$ term by 4. Then, multiply both sides of the equation by $y$ to make it easier to work with; $4(\frac{x}{y} + 1) = 10$, $\frac{4x}{y} + 4 = 10$, $4x + 4y = 10y$, $4x = 6y$, $y = \frac{2}{3}x$.

**16. e.** Isolate $g$ on one side of the equation; $fg + 2f - g = 2 - (f + g)$, $fg + 2f - g = 2 - f - g$, $fg = 2 - 3f$, $g = \frac{2 - 3f}{f}$.

**17. d.** Isolate $b$ on one side of the equation; $a(3a) - b(4 + a) = -(a^2 + ab)$, $3a^2 - 4b - ab = -a^2 - ab$, $4a^2 - 4b = 0$, $4a^2 = 4b$, $b = a^2$.

**18. b.** Isolate $g$ on one side of the equation; $4g^2 - 1 = 16h^2 - 1$, $4g^2 = 16h^2$, $g^2 = 4h^2$, $g = 2h$.

**19. a.** Isolate $x$ on one side of the equation; $8x^2 - 4y^2 + x^2 = 0$, $9x^2 - 4y^2 = 0$, $9x^2 = 4y^2$, $x^2 = \frac{4}{9}y^2$, $x = \frac{2}{3}y$.

**20. b.** Isolate $y$ on one side of the equation; $\frac{10(x^2y)}{(xy^2)} = 5y$, $10\frac{x}{y} = 5y$, $10x = 5y^2$, $2x = y^2$, $y = \sqrt{2x}$.

## ► Chapter 2

**1. e.** To solve the equation, add 12 to both sides of the equation: $a - 12 = 12$, $a - 12 + 12 = 12 + 12$, $a = 24$.

**2. c.** To solve the inequality, divide both sides of the inequality by 6: $6p \geq 10$, $\frac{6p}{6} \geq \frac{10}{6}$, $p \geq \frac{5}{3}$.

**3. a.** To solve the equation, subtract 10 from both sides of the equation: $x + 10 = 5$, $x + 10 - 10 = 5 - 10$, $x = -5$.

**4. e.** To solve the equation, multiply both sides of the equation by 8: $\frac{k}{8} = 8$, $(8)\frac{k}{8} = (8)(8)$, $k = 64$.

**5. d.** To solve the inequality, divide both sides of the inequality by $-3$: $-3n < 12$, $\frac{-3n}{-3} > \frac{12}{-3}$. Remember, when multiplying or dividing both sides of an inequality by a negative number, you must reverse the inequality symbol; $n > -4$.

**6. c.** To solve the equation, subtract 5 from both sides of the equation, then divide by 9: $9a + 5 = -22$, $9a + 5 - 5 = -22 - 5$, $9a = -27$, $a = -3$.

**7. a.** First, multiply $(x + 2)$ by 4: $4(x + 2) = 4x + 8$. Then, subtract $3x$ from both sides of the inequality and subtract 8 from both sides of the inequality:

$$3x - 6 \le 4x + 8$$
$$3x - 6 - 3x \le 4x + 8 - 3x$$
$$-6 \le x + 8$$
$$-6 - 8 \le x + 8 - 8$$
$$x \ge -14$$

**8. d.** First, combine like terms on each side of the equation; $6x - 4x = 2x$ and $4 - 9 = -5$. Now, subtract $2x$ from both sides of the equation and add 5 to both sides of the equation:

$$2x + 9 = 6x - 5$$
$$2x - 2x + 9 = 6x - 2x - 5$$
$$9 = 4x - 5$$
$$9 + 5 = 4x - 5 + 5$$
$$14 = 4x$$

Finally, divide both sides of the equation by 4: $\frac{14}{4} = \frac{4x}{4}$, $x = \frac{14}{4} = \frac{7}{2}$.

**9. a.** First, multiply $(x + 3)$ by $-8$ and multiply $(-2x + 10)$ by 2: $-8(x + 3) = -8x - 24$, $2(-2x + 10) = -4x + 20$. Then, add $8x$ to both sides of the inequality and subtract 20 from both sides of the inequality:

$$-8x - 24 \le -4x + 20$$
$$-8x - 24 + 8x \le -4x + 20 + 8x$$
$$-24 \le 4x + 20$$
$$-24 - 20 \le 4x + 20 - 20$$
$$-44 \le 4x$$

Finally, divide both sides of the inequality by 4: $-\frac{44}{4} \le \frac{4x}{4}$, $x \ge -11$.

**10. e.** First, reduce the fraction $\frac{(3c^2)}{(6c)}$ by dividing the numerator and denominator by $3c$; $\frac{(3c^2)}{(6c)} = \frac{c}{2}$. Now, subtract 9 from both sides of the equation and then multiply both sides of the equation by 2:

$$\frac{c}{2} + 9 = 15$$
$$\frac{c}{2} + 9 - 9 = 15 - 9$$
$$\frac{c}{2} = 6$$
$$(2)(\tfrac{c}{2}) = (6)(2)$$
$$c = 12$$

**11. b.** Cross multiply and solve for $w$: $(w)(18) = (-6)(w + 8)$, $18w = -6w - 48$, $24w = -48$, $w = -2$.

**12. d.** Cross multiply and solve for $x$: $(10x)(3) = (7)(5x - 10)$, $30x = 35x - 70$, $-5x = -70$, $x = 14$.

**13. d.** Cross multiply and solve for $a$: $(4a + 4)(4) = (7)(-2 + 3a)$, $16a + 16 = -14 + 21a$, $16a + 30 = 21a$, $5a = 30$, $a = 6$.

**14. a.** Cross multiply and solve for $y$: $(6)(-2y - 3) = (10)(-y - 1)$, $-12y - 18 = -10y - 10$, $-2y - 18 = -10$, $-2y = 8$, $y = -4$.

**15. c.** First, reduce $\frac{5g}{g}$ by canceling $g$ from the numerator and denominator: $\frac{5g}{g} = \frac{5}{1}$. Now, cross multiply and solve for $g$: $(5)(g - 1) = (1)(g + 7)$, $5g - 5 = g + 7$, $4g - 5 = 7$, $4g = 12$, $g = 3$.

**16. e.** Add $x$ to both sides of the equation and subtract 6 from both sides of the equation; $\frac{1}{2}x + 6 = -x - 3$, $\frac{3}{2}x = -9$. Multiply both sides of the equation by $\frac{2}{3}$ to isolate $x$: $(\frac{2}{3})(\frac{3}{2}x) = -9(\frac{2}{3})$, $x = -6$. Since $x = -6$, $-2x = -2(-6) = 12$.

**17. b.** First, cross multiply: $(2)(3x - 8) = 9x + 5$, $6x - 16 = 9x + 5$. Subtract $6x$ from both sides of the equation and subtract 5 from both sides of the equation, $6x - 16 = 9x + 5$, $3x = -21$. Divide by 3 to solve for $x$: $\frac{3x}{3} = \frac{-21}{3}$, $x = -7$. Since $x = -7$, $x + 7 = -7 + 7 = 0$.

**18. a.** Subtract $\frac{8}{3}x$ and $\frac{8}{3}$ from both sides of the equation: $9x + \frac{8}{3} = \frac{8}{3}x + 9$, $\frac{19}{3}x = \frac{19}{3}$. Multiply both sides of the equation by $\frac{3}{19}$: $(\frac{3}{19})(\frac{19}{3}x) = (\frac{19}{3})(\frac{3}{19})$, $x = 1$. Since $x = 1$, $\frac{3}{8}x = \frac{3}{8}(1) = \frac{3}{8}$.

**19. d.** Although you could solve the first equation for $x$ and substitute that value into the given expression, look at the relationship between the equation and the expression; $4x + 2$ is exactly half of $8x + 4$. Therefore, the value of $4x + 2$ will be half the value of $8x + 4$; $\frac{14}{2} = 7$; $4x + 2 = 7$.

**20. d.** Although you could solve the first equation for $c$ and substitute that value into the given expression, look at the relationship between the equation and the expression. $33c - 21$ is exactly three times $11c - 7$. Therefore, the value of $33c - 21$ will be three times the value of $11c - 7$; $3(8) = 24$; $33c - 21 = 24$.

**21. e.** A fraction is undefined when its denominator is equal to 0. Set $x - 8$ equal to 0 and solve for $x$; $x - 8 = 0$, $x = 8$. The fraction is undefined when $x = 8$.

**22. c.** A fraction is undefined when its denominator is equal to 0. Set $6d$ equal to 0 and solve for $d$; $6d = 0$, $d = 0$. The fraction is undefined when $d = 0$.

**23. b.** A fraction is undefined when its denominator is equal to 0. Set $6a + 18 - 4a$ equal to 0 and solve for $a$; $6a + 18 - 4a = 0$, $2a + 18 = 0$, $2a = -18$, $a = -9$. The fraction is undefined when $a = -9$.

**24. c.** A fraction is undefined when its denominator is equal to 0. Set $8 - 8y + 4$ equal to 0 and solve for $y$; $8 - 8y + 4 = 0$, $-8y + 12 = 0$, $-8y = -12$, $y = \frac{3}{2}$. The fraction is undefined when $y = \frac{3}{2}$.

**25. b.** A fraction is undefined when its denominator is equal to 0. Set each term of the denominator equal to 0 and solve for $x$; $2x(x - 5) = 0$; $2x = 0$, $x = 0$ and $x - 5 = 0$, $x = 5$. The fraction is undefined when $x = 0$ or 5.

**26. b.** Create an equation that describes the situation. If $x$ represents the number, then 3 times the number is $3x$. Five less than that is $3x - 5$. Set that expression equal to 10, and solve for $x$: $3x - 5 = 10$, $3x = 15$, $x = 5$.

**27. d.** Create an equation that describes the situation. If $x$ represents the number, then one-fourth the number is $\frac{1}{4}x$. Three more than that is $\frac{1}{4}x + 3$. Three less than the number is $x - 3$. Set these expressions equal to each other, and solve for $x$: $\frac{1}{4}x + 3 = x - 3$, $3 = \frac{3}{4}x - 3$, $6 = \frac{3}{4}x$, $x = 8$.

**28. e.** Create an equation that describes the situation. If $x$ represents the first integer, $x + 1$ represents the second integer, and $x + 2$ represents the third integer. The sum of the integers is $x + x + 1 + x + 2 = 3x + 3$. Set this expression equal to the sum of the integers, 63, and solve for $x$: $3x + 3 = 63$, $3x = 60$, $x = 20$. Since $x$ is the first integer, 21 is the second integer, and 22 is the third (and largest) integer.

**29. a.** Create an equation that describes the situation. If $x$ represents the first odd whole number, $x + 2$ represents the second, $x + 4$ represents the third, and $x + 6$ represents the fourth. The sum of the numbers is $x + x + 2 + x + 4 + x + 6 = 4x + 12$. Set this expression equal to the sum of the numbers, 48, and solve for $x$: $4x + 12 = 48$, $4x = 36$, $x = 9$. Since $x$ is the first number, it is the smallest of the four consecutive, odd whole numbers.

**30. c.** Create an equation that describes the situation. If $x$ represents the first even integer, $x + 2$ represents the second integer and $x + 4$ represents the third. The sum of the integers is $x + x + 2 + x + 4 = 3x + 6$. Set this expression equal to the sum of the integers, $-18$, and solve for $x$: $3x + 6 = -18$, $3x = -24$, $x = -8$. Since $x$ is the first number, it is the smallest of the three consecutive, even integers.

## ▶ Chapter 3

**1. c.** To find the product of two binomials, multiply the first term of each binomial, the outside terms, the inside terms, and the last terms. Then, add the products; $(x - 3)(x + 7) = x^2 + 7x - 3x - 21 = x^2 + 4x - 21$.

**2. d.** To find the product of two binomials, multiply the first term of each binomial, the outside terms, the inside terms, and the last terms. Then, add the products; $(x - 6)(x - 6) = x^2 - 6x - 6x + 36 = x^2 - 12x + 36$.

**3. a.** To find the product of two binomials, multiply the first term of each binomial, the outside terms, the inside terms, and the last terms. Then, add the products; $(x-1)(x+1) = x^2 + x - x - 1 = x^2 - 1$.

**4. e.** $(x+c)^2 = (x+c)(x+c)$. To find the product of two binomials, multiply the first term of each binomial, the outside terms, the inside terms, and the last terms. Then, add the products; $(x+c)(x+c) = x^2 + cx + cx + c^2 = x^2 + 2cx + c^2$.

**5. b.** To find the product of two binomials, multiply the first term of each binomial, the outside terms, the inside terms, and the last terms. Then, add the products; $(2x+6)(3x-9) = 6x^2 - 18x + 18x - 54 = 6x^2 - 54$.

**6. b.** To find the factors of a quadratic, begin by finding two numbers whose product is equal to the constant of the quadratic. Of those numbers, find the pair that adds to the coefficient of the $x$ term of the quadratic; $-3$ and $2$ multiply to $-6$ and add to $-1$. Therefore, the factors of $x^2 - x - 6$ are $(x-3)$ and $(x+2)$.

**7. d.** To find the factors of a quadratic, begin by finding two numbers whose product is equal to the constant of the quadratic. Of those numbers, find the pair that adds to the coefficient of the $x$ term of the quadratic. This quadratic has no $x$ term—the sum of the products of the outside and inside terms of the factors is 0; $-2$ and $2$ multiply to $-4$ and add to 0. Therefore, the factors of $x^2 - 4$ are $(x-2)$ and $(x+2)$.

**8. a.** To find the factors of a quadratic, begin by finding two numbers whose product is equal to the constant of the quadratic. Of those numbers, find the pair that adds to the coefficient of the $x$ term of the quadratic; $-4$ and $-7$ multiply to 28 and add to $-11$. Therefore, the factors of $x^2 - 11x + 28$ are $(x-4)$ and $(x-7)$.

**9. e.** The roots of a quadratic are the solutions of the quadratic. Factor the quadratic and set each factor equal to 0 to find the roots. To find the factors of a quadratic, begin by finding two numbers whose product is equal to the constant of the quadratic. Of those numbers, find the pair that adds to the coefficient of the $x$ term of the quadratic; $-2$ and $-16$ multiply to 32 and add to $-18$. Therefore, the factors of $x^2 - 18x + 32$ are $(x-2)$ and $(x-16)$. Set each factor equal to 0 and solve for $x$: $x - 2 = 0$, $x = 2$, and $x - 16 = 0$, $x = 16$. The roots of $x^2 - 18x + 32$ are 2 and 16.

**10. c.** The roots of a quadratic are the solutions of the quadratic. Factor the quadratic and set each factor equal to 0 to find the roots. To find the factors of a quadratic, begin by finding two numbers whose product is equal to the constant of the quadratic. Of those numbers, find the pair that adds to the coefficient of the $x$ term of the quadratic; $-4$ and 12 multiply to $-48$ and add to 8. Therefore, the factors of $x^2 + 8x - 48$ are $(x-4)$ and $(x+12)$. Set each factor equal to 0 and solve for $x$: $x - 4 = 0$, $x = 4$, and $x + 12 = 0$, $x = -12$. The roots of $x^2 + 8x - 48$ are 4 and $-12$.

**11. a.** Factor the denominator. To find the factors of a quadratic, begin by finding two numbers whose product is equal to the constant of the quadratic. Of those numbers, find the pair that adds to the coefficient of the $x$ term of the quadratic. This quadratic has no $x$ term—the sum of the products of the outside and inside terms of the factors is 0; $-9$ and 9 multiply to $-81$ and add to 0. Therefore, the factors of $x^2 - 81$ are $(x-9)$ and $(x+9)$. Cancel the $(x+9)$ terms from the numerator and denominator, leaving 1 in the numerator and $(x-9)$ in the denominator.

**12. b.** Factor the numerator and denominator. The numerator factors into $(x-8)(x+2)$ and the denominator factors into $(x-8)(x+7)$. Cancel the $(x-8)$ terms from the numerator and denominator, leaving $(x+2)$ in the numerator and $(x+7)$ in the denominator.

**13. c.** Factor the numerator and denominator. The numerator factors into $(x+5)(x-9)$ and the denominator factors into $(x+5)(x+6)$. Cancel the $(x+5)$ terms from the numerator and denominator, leaving $(x - 9)$ in the numerator and $(x + 6)$ in the denominator.

**14. b.** Factor the numerator and denominator. The numerator factors into $(x + 11)(x + 3)$ and the denominator factors into $(x + 11)(x - 3)$. Cancel the $(x + 11)$ terms from the numerator and denominator, leaving $(x + 3)$ in the numerator and $(x - 3)$ in the denominator.

**15. e.** Combine like terms on one side of the equation and set the expression equal to 0; $x^2 - x = 12$, $x^2 - x - 12 = 0$. Factor the quadratic and set each factor equal to 0 to find the solutions for $x$. To find the factors of a quadratic, begin by finding two numbers whose product is equal to the constant of the quadratic. Of those numbers, find the pair that adds to the coefficient of the $x$ term of the quadratic; $-4$ and 3 multiply to $-12$ and add to $-1$. Therefore, the factors of $x^2 - x - 12$ are $(x - 4)$ and $(x + 3)$. Set each factor equal to 0 and solve for $x$: $x - 4 = 0, x = 4$, and $x + 3 = 0, x = -3$; $x^2 - x = 12$ when $x$ equals 4 or $-3$. Trial and error (plugging each answer choice into the equation $x^2 - x = 12$) could be used, but only on a multiple-choice SAT question. It is important to be able to solve questions like this that could occur in the grid-in section of the SAT.

**16. d.** Combine like terms on one side of the equation and set the expression equal to 0; $x^2 - 3x - 30 = 10$, $x^2 - 3x - 40 = 0$. Factor the quadratic and set each factor equal to 0 to find the solutions for $x$. To find the factors of a quadratic, begin by finding two numbers whose product is equal to the constant of the quadratic. Of those numbers, find the pair that adds to the coefficient of the $x$ term of the quadratic; $-8$ and 5 multiply to $-40$ and add to $-3$. Therefore, the factors of $x^2 - 3x - 40$ are $(x - 8)$ and $(x + 5)$. Set each factor equal to 0 and solve for $x$: $x - 8 = 0, x = 8$, and $x + 5 = 0, x = -5$; $x^2 - 3x - 30 = 10$ when $x$ equals 8 or $-5$.

**17. e.** Cross multiply: $(x + 5)(x + 4) = (4)(6x - 6)$, $x^2 + 9x + 20 = 24x - 24$. Combine like terms on one side of the equation and set the expression equal to 0; $x^2 + 9x + 20 = 24x - 24$, $x^2 - 15x + 44 = 0$. Factor the quadratic and set each factor equal to 0 to find the solutions for $x$. To find the factors of a quadratic, begin by finding two numbers whose product is equal to the constant of the quadratic. Of those numbers, find the pair that adds to the coefficient of the $x$ term of the quadratic; $-4$ and $-11$ multiply to 44 and add to $-15$. Therefore, the factors of $x^2 - 15x + 44$ are $(x - 4)$ and $(x - 11)$. Set each factor equal to 0 and solve for $x$: $x - 4 = 0, x = 4$, and $x - 11 = 0, x = 11$.

**18. c.** Use FOIL to multiply the binomials: $(x - 2)(x + 6) = x^2 + 6x - 2x - 12 = x^2 + 4x - 12$. Combine like terms on one side of the equation and set the expression equal to 0; $x^2 + 4x - 12 = -16$, $x^2 + 4x + 4 = 0$. Factor the quadratic and set each factor equal to 0 to find the solutions for $x$. To find the factors of a quadratic, begin by finding two numbers whose product is equal to the constant of the quadratic. Of those numbers, find the pair that adds to the coefficient of the $x$ term of the quadratic; 2 and 2 multiply to 4 and add to 4. Therefore, the factors of $x^2 + 4x + 4$ are $(x + 2)$ and $(x + 2)$. Since both factors are the same, set either factor equal to 0 and solve for $x$: $x + 2 = 0, x = -2$.

**19. d.** Use FOIL to multiply the binomials: $(x - 7)(x - 5) = x^2 - 5x - 7x + 35 = x^2 - 12x + 35$. Combine like terms on one side of the equation and set the expression equal to 0; $x^2 - 12x + 35 = -1$, $x^2 - 12x + 36 = 0$. Factor the quadratic and set each factor equal to 0 to find the solutions for $x$. To find the factors of a quadratic, begin by finding two numbers whose product is equal to the constant of the quadratic. Of those numbers, find the pair that adds to the coefficient of the $x$ term of the quadratic; $-6$ and $-6$ multiply to 36 and add to $-12$. Therefore, the factors of $x^2 - 12x + 36$ are $(x - 6)$ and $(x - 6)$. Since both factors are the same, set either factor equal to 0 and solve for $x$: $x - 6 = 0, x = 6$.

**20. a.** Write an algebraic equation that describes the situation. If $x$ is the number, then $x^2$ is the square of the number. Two less than three times the number is $3x - 2$. Since the square of the number is equal to two less than three times the number, $x^2 = 3x - 2$. Combine like terms on one side of the equation and set the expression equal to 0; $x^2 = 3x - 2$, $x^2 - 3x + 2 = 0$. Factor the quadratic and set each factor equal to 0 to find the solutions for $x$. To find the factors of a quadratic, begin by finding two numbers whose product is equal to the constant of the quadratic. Of those numbers, find the pair that adds to the coefficient of the $x$ term of the quadratic; $-1$ and $-2$ multiply to 2 and add to $-3$. Therefore, the factors of $x^2 - 3x + 2$ are $(x - 1)$ and $(x - 2)$. Set each factor equal to 0 and solve for $x$: $x - 1 = 0$, $x = 1$, and $x - 2 = 0$, $x = 2$.

**21. a.** A fraction is undefined when its denominator is equal to 0. Factor the quadratic in the denominator and set each factor equal to 0 to find the values that make the fraction undefined. To find the factors of a quadratic, begin by finding two numbers whose product is equal to the constant of the quadratic. Of those numbers, find the pair that adds to the coefficient of the $x$ term of the quadratic; $-6$ and 7 multiply to $-42$ and add to 1. Therefore, the factors of $x^2 + x - 42$ are $(x - 6)$ and $(x + 7)$. Set each factor equal to 0 and solve for $x$: $x - 6 = 0$, $x = 6$, and $x + 7 = 0$, $x = -7$. When $x$ equals 6 or $-7$, the fraction is undefined.

**22. d.** A fraction is undefined when its denominator is equal to 0. Factor the quadratic in the denominator and set each factor equal to 0 to find the values that make the fraction undefined. To find the factors of a quadratic, begin by finding two numbers whose product is equal to the constant of the quadratic. Of those numbers, find the pair that adds to the coefficient of the $x$ term of the quadratic; $-1$ and $-7$ multiply to 7 and add to $-8$. Therefore, the factors of $x^2 - 8x + 7$ are $(x - 1)$ and $(x - 7)$. Set each factor equal to 0 and solve for $x$: $x - 1 = 0$, $x = 1$, and $x - 7 = 0$, $x = 7$. When $x$ equals 1 or 7, the fraction is undefined.

**23. b.** A fraction is undefined when its denominator is equal to 0. Set the denominator equal to 0 and solve for $x$; $x^2 - 16 = 0$, $x^2 = 16$, $x = -4$ or 4.

**24. c.** A fraction is undefined when its denominator is equal to 0. Set the denominator equal to 0 and solve for $x$; $9x^2 - 1 = 0$, $9x^2 = 1$, $x^2 = \frac{1}{9}$, $x = -\frac{1}{3}$ or $\frac{1}{3}$.

**25. d.** A fraction is undefined when its denominator is equal to 0. Factor the quadratic in the denominator and set each factor equal to 0 to find the values that make the fraction undefined. To find the factors of a quadratic, begin by finding two numbers whose product is equal to the constant of the quadratic. Since the first term of the quadratic is $2x^2$, the factors of the quadratic will be $(2x + c)(x + d)$, where $c$ and $d$ are constants that multiply to 72 and add to $-25$ after either $c$ or $d$ is multiplied by 2. $-8$ and $-9$ multiply to 72, and $2(8) + 9 = 25$. Therefore, the factors of $2x^2 - 25x + 72$ are $(2x - 9)$ and $(x - 8)$. Set each factor equal to 0 and solve for $x$: $2x - 9 = 0$, $2x = 9$, $x = \frac{9}{2}$, and $x - 8 = 0$, $x = 8$. When $x$ equals $\frac{9}{2}$ or 8, the fraction is undefined.

**26. d.** A parabola of the form $y = x^2 + c$ has its vertex at $(0,c)$. Therefore, the vertex of this parabola is at $(0,4)$. This parabola is similar to the parabola $y = x^2$, but shifted up 4 units.

**27. e.** A parabola of the form $y = (x + c)^2 + d$ has its vertex at $(-c,d)$. Therefore, the equation of a parabola whose vertex is $(-3,-4)$ is $y = (x + 3)^2 - 4$. This parabola is similar to the parabola $y = x^2$, but shifted left 3 units and down 4 units.

**28. b.** A parabola of the form $y = (x + c)^2 + d$ has its vertex at $(-c,d)$. Therefore, the vertex of the parabola whose equation is $y = (x + 2)^2 + 2$ is $(-2,2)$. This parabola is similar to the parabola $y = x^2$, but shifted left 2 units and up 2 units.

**29. c.** A parabola of the form $y = (x + c)^2 + d$ has its vertex at $(-c,d)$. Therefore, the equation of a parabola whose vertex is $(5,0)$ is $y = (x-5)^2 + 0$, or $y = (x-5)^2$. This parabola is similar to the parabola $y = x^2$, but shifted right 5 units.

**30. a.** The vertex of this parabola is at $(1,-2)$. A parabola of the form $y = (x+c)^2 + d$ has its vertex at $(-c,d)$. Therefore, the equation of a parabola whose vertex is $(1,-2)$ is $y = (x-1)^2 - 2$. This parabola is similar to the parabola $y = x^2$, but shifted right 1 unit and down 2 units.

## ▶ Chapter 4

**1. e.** Begin by multiplying the first two terms: $-3x(x+6) = -3x^2 - 18x$. Multiply $(-3x^2 - 18x)$ by $(x-9)$: $(-3x^2 - 18x)(x-9) = -3x^3 + 27x^2 - 18x^2 + 162x = -3x^3 + 9x^2 + 162x$.

**2. d.** Multiply each term of the trinomial by each term of the binomial: $(x^2)(x) = x^3$, $(5x)(x) = 5x^2$, $(-7)(x) = -7x$, $(x^2)(2) = 2x^2$, $(5x)(2) = 10x$, $(-7)(2) = -14$. Add the products and combine like terms: $x^3 + 5x^2 + -7x + 2x^2 + 10x + -14 = x^3 + 7x^2 + 3x - 14$.

**3. e.** Begin by multiplying the first two terms: $(x-6)(x-3) = x^2 - 3x - 6x + 18 = x^2 - 9x + 18$. Multiply $(x^2 - 9x + 18)$ by $(x-1)$: $(x^2 - 9x + 18)(x-1) = x^3 - 9x^2 + 18x - x^2 + 9x - 18 = x^3 - 10x^2 + 27x - 18$.

**4. e.** 16 is the largest constant common to $64x^3$ and $16x$, and $x$ is the largest common variable. Factor out $16x$ from both terms: $\frac{64x^3}{16x} = 4x^2$ and $\frac{-16x}{16x} = -1$. $64x^3 - 16x = 16x(4x^2 - 1)$. Next, factor $4x^2 - 1$; $(2x)(2x) = 4x^2$, and $(1)(-1) = -1$. $4x^2 - 1 = (2x-1)(2x+1)$, so the factors of $64x^3 - 16x$ are $16x(2x-1)(2x+1)$.

**5. b.** The largest constant common to each term is 2, and $x$ is the largest common variable. Factor out $2x$ from every term: $2x^3 + 8x^2 - 192x$: $2x(x^2 + 4x - 96)$. Factor $x^2 + 4x - 96$ into $(x-8)(x+12)$. The factors of $2x^3 + 8x^2 - 192x$ are $2x(x-8)(x+12)$.

**6. c.** First, multiply the terms on the left side of the equation. $x(x-1) = x^2 - x$, $(x^2 - x)(x+1) = x^3 + x^2 - x^2 - x = x^3 - x$. Therefore, $x^3 - x = 27 - x$. Add $x$ to both sides of the equation; $x^3 - x + x = 27 - x + x$, $x^3 = 27$. The cube root of 27 is 3, so the root, or solution, of $x(x-1)(x+1) = 27 - x$ is $x = 3$.

**7. a.** Factor the numerator and denominator; $x^2 + 8x = x(x+8)$; $x^3 - 64x = x(x^2 - 64) = x(x-8)(x+8)$. Cancel the $x$ terms and the $(x+8)$ terms in the numerator and denominator, leaving 1 in the numerator and $(x-8)$ in the denominator.

**8. c.** Factor the numerator and denominator; $x^2 + 6x + 5 = (x+1)(x+5)$. $x^3 - 25x = x(x^2 - 25) = x(x-5)(x+5)$. Cancel the $(x+5)$ terms in the numerator and denominator, leaving $(x+1)$ in the numerator and $x(x-5) = (x^2 - 5x)$ in the denominator.

**9. c.** Factor the numerator and denominator; $2x^2 + 4x = 2x(x+2)$; $4x^3 - 16x^2 - 48x = 4x(x^2 - 4x - 12) = 4x(x-6)(x+2)$. Cancel the $2x$ term in the numerator with the $4x$ term in denominator, leaving 2 in the denominator. Cancel the $(x+2)$ terms in the numerator and denominator, leaving $2(x-6) = 2x - 12$ in the denominator.

**10. b.** A fraction is undefined when its denominator is equal to 0. Set the denominator equal to 0 and solve for $x$; $x^3 + 125 = 0$, $x^3 = -125$, $x = -5$.

**11. b.** A fraction is undefined when its denominator is equal to 0. Factor the polynomial in the denominator and set each factor equal to 0 to find the values that make the fraction undefined; $x^3 + 3x^2 - 4x = x(x+4)(x-1)$; $x = 0$; $x + 4 = 0$, $x = -4$; $x - 1 = 0$, $x = 1$. The fraction is undefined when $x$ is equal to $-4$, 0, or 1.

**12. a.** A fraction is undefined when its denominator is equal to 0. Factor the polynomial in the denominator and set each factor equal to 0 to find the values that make the fraction undefined; $4x^3 + 44x^2 + 120x = 4x(x^2 + 11x + 30) =$

$4x(x + 5)(x + 6)$; $4x = 0$, $x = 0$; $x + 5 = 0$, $x = -5$; $x + 6 = 0$, $x = -6$. The fraction is undefined when $x$ is equal to $-6$, $-5$, or $0$.

**13. d.** If the number is $x$, then the cube of the number is $x^3$. Twice the square of the number is $2x^2$. The difference in those values is equal to 80 times the number ($80x$). Therefore, $x^3 - 2x^2 = 80x$. Move all terms onto one side of the equation, and factor the polynomial; $x^3 - 2x^2 = 80x$, $x^3 - 2x^2 - 80x = 0$; $x^3 - 2x^2 - 80x = x(x^2 - 2x - 80) = x(x + 8)(x - 10)$. Set each factor equal to 0 to find the values of $x$ that make the equation true. $x = 0$; $x + 8 = 0$, $x = -8$; $x - 10 = 0$, $x = 10$. The given situation is true for the numbers 0, $-8$, and 10, but only 10 is greater than 0. Trial and error (plugging each answer choice into the equation $x^3 - 2x^2 = 80x$) could be used, but only on a multiple-choice SAT question. It is important to be able to solve questions like this that could occur in the grid-in section.

**14. a.** If the number is $x$, then four times the cube of the number is $4x^3$. That value is equal to 48 times the number ($48x$) minus four times the square of the number ($4x^2$). Therefore, $4x^3 = 48x - 4x^2$. Move all terms onto one side of the equation, and factor the polynomial; $4x^3 = 48x - 4x^2$, $4x^3 + 4x^2 - 48x = 0$; $4x^3 + 4x^2 - 48x = 4x(x^2 + x - 12) = 4x(x + 4)(x - 3)$. Set each factor equal to 0 to find the values of $x$ that make the equation true. $4x = 0$, $x = 0$; $x + 4 = 0$, $x = -4$; $x - 3 = 0$, $x = 3$. The given situation is true for the numbers 0, $-4$, and 3, but only 3 is greater than 0. Trial and error could also be used for a multiple-choice problem such as this.

**15. d.** If the first integer is $x$, then the second integer is $(x + 1)$ and the third integer is $(x + 2)$. The product of these integers is $x(x + 1)(x + 2)$; $x(x + 1) = x^2 + x$; $(x^2 + x)(x + 2) = x^3 + 2x^2 + x^2 + 2x = x^3 + 3x^2 + 2x$. This polynomial is equal to the cube of the first integer, $x^3$, plus 56. Therefore, $x^3 + 3x^2 + 2x = x^3 + 56$. Place all terms on one side of the equation, combining like terms; $x^3 + 3x^2 + 2x = x^3 + 56$, $3x^2 + 2x - 56 = 0$. Factor the polynomial: $3x^2 + 2x - 56 = (3x + 14)(x - 4)$. Set each term equal to 0 to find the values of $x$ that make the equation true; $3x + 14 = 0$, $3x = -14$, $x = -\frac{14}{3}$, but you are looking for a positive integer; $x - 4 = 0$, $x = 4$. If 4 is the first of the consecutive integers, then 5 is the second integer and 6 is the third, and largest, integer. Again, trial and error could be used to find the solution.

## ▶ Chapter 5

**1. b.** Find the square root of the coefficient and the variable. $\sqrt{(32x^2)} = \sqrt{32}\sqrt{(x^2)} = x\sqrt{32}$. Next, factor $\sqrt{32}$ into two radicals, one of which is a perfect square. $\sqrt{32} = (\sqrt{16})(\sqrt{2}) = 4\sqrt{2}$. Therefore, $\sqrt{(32x^2)} = 4x\sqrt{2}$.

**2. a.** Factor $\sqrt{(a^3)}$ into two radicals; $a^2$ is a perfect square, so factor $\sqrt{(a^3)}$ into $\sqrt{a}\sqrt{(a^2)} = a\sqrt{a}$. Multiply the coefficient of the given expression by $a\sqrt{a}$: $(a^3)(a\sqrt{a}) = a^4\sqrt{a}$.

**3. a.** Factor $\sqrt{4g}$ into two radicals; 4 is a perfect square, so factor $\sqrt{4g}$ into $\sqrt{4}\sqrt{g} = 2\sqrt{g}$. Simplify the fraction by dividing the numerator by the denominator. Cancel the $\sqrt{g}$ terms from the numerator and denominator. That leaves $\frac{4}{2} = 2$.

**4. a.** The cube root of $27y^3 = 3y$, since $(3y)(3y)(3y) = 27y^3$. Factor the denominator into two radicals. $\sqrt{(27y^2)} = (\sqrt{(9y^2)})(\sqrt{3})$. The square root of $9y^2 = 3y$, since $(3y)(3y) = 9y^2$. The expression is now equal to $\frac{3y}{3y\sqrt{3}}$. Cancel the $3y$ terms from the numerator and denominator, leaving $\frac{1}{\sqrt{3}}$. Simplify the fraction by multiplying the numerator and denominator by $\sqrt{3}$: $(\frac{1}{\sqrt{3}})(\frac{\sqrt{3}}{\sqrt{3}}) = \frac{\sqrt{3}}{3}$.

**5. c.** Factor each term in the numerator: $\sqrt{(a^2b)} = \sqrt{(a^2)}(\sqrt{b}) = a\sqrt{b}$; $\sqrt{(a^2b)} = (\sqrt{a})(\sqrt{(b^2)}) = b\sqrt{a}$. Next, multiply the two radicals. Multiply the coefficients of each radical and multiply the radicands of each radical: $(a\sqrt{b})(b\sqrt{a}) = ab\sqrt{ab}$. The expression is now $\frac{ab\sqrt{ab}}{\sqrt{ab}}$. Cancel the $\sqrt{ab}$ terms from the numerator and denominator, leaving $ab$.

**6. b.** A term with a negative exponent can be rewritten as the reciprocal of the term with a positive exponent. $(\sqrt{\frac{m^3}{n^5}})^{-2} = (\frac{1}{\sqrt{\frac{m^3}{n^5}}})^2$. Square the numerator and denominator. $(1)^2 = 1$, $(\sqrt{\frac{m^3}{n^5}})^2 = \sqrt{\frac{m^3}{n^5}}$. Therefore, $(\frac{1}{\sqrt{\frac{m^3}{n^5}}})^2 = (\frac{1}{\sqrt{\frac{m^3}{n^5}}}) = \frac{n^5}{m^3}$.

**7. d.** First, cube the $ab$ term; $(ab)^3 = a^3b^3$. Next, raise the fraction $\frac{(a^3b^3)}{b}$ to the fourth power. Multiply each exponent of the $a$ and $b$ terms by 4. $(\frac{(a^3b^3)}{b})^4 = \frac{(a^{12}b^{12})}{b^4}$. To divide $b^{12}$ by $b^4$, subtract the exponents; $\frac{b^{12}}{b^4} = b^8$. Therefore, $\frac{(a^{12}b^{12})}{b^4} = a^{12}b^8$.

**8. c.** First, cube the $4g^2$ term. Cube the constant 4 and multiply the exponent of $g$ (2) by 3: $(4g^2)^3 = 64g^6$. Next, multiply $64g^6$ by $g^4$. Add the exponents of the $g$ terms. $(64g^6)(g^4) = 64g^{10}$. Finally, take the square root of $64g^{10}$; $(64g^{10})^{\frac{1}{2}} = 8g^5$, since $(8g^5)(8g^5) = 64g^{10}$.

**9. e.** First, find the square root of $9pr$; $\sqrt{9pr} = \sqrt{9}\sqrt{pr} = 3\sqrt{pr}$. The denominator $(pr)^{-\frac{3}{2}}$ has a negative exponent, so it can be rewritten in the numerator with a positive exponent; $\sqrt{pr}$ can be written as $(pr)^{\frac{1}{2}}$, since a value raised to the exponent $\frac{1}{2}$ is another way of representing the square root of the value. The expression is now $3(pr)^{\frac{1}{2}}(pr)^{\frac{3}{2}}$. To multiply the $pr$ terms, add the exponents; $\frac{1}{2} + \frac{3}{2} = \frac{4}{2} = 2$, so $3(pr)^{\frac{1}{2}}(pr)^{\frac{3}{2}} = 3(pr)^2 = 3p^2r^2$.

**10. b.** First, square $\frac{x}{y}$: $(\frac{x}{y})^2 = \frac{x^2}{y^2}$. Next, look at the $(\frac{y}{x})^{-2}$ term. A fraction with a negative exponent can be rewritten as the reciprocal of the fraction with a positive exponent. $(\frac{y}{x})^{-2} = (\frac{x}{y})^2 = \frac{x^2}{y^2}$. Multiply the fractions in the numerator by adding the exponents of the fractions: $(\frac{x^2}{y^2})(\frac{x^2}{y^2}) = (\frac{x^4}{y^4})$. Finally, divide this fraction by $xy$; $\frac{(\frac{x^4}{y^4})}{xy} = (\frac{x^4}{y^4})(\frac{1}{xy}) = \frac{x^4}{xy^5} = \frac{x^3}{y^5}$.

**11. e.** Since $(a^{\frac{2}{3}})^2 = a^{\frac{4}{3}}$, the value of $a^{\frac{4}{3}}$ is equal to the value of $a^{\frac{2}{3}}$ squared. Therefore, $a^{\frac{4}{3}} = 6^2 = 36$.

**12. d.** $(\sqrt{p})^4 = (p^{\frac{1}{2}})^4$. Multiply the exponents: $(p^{\frac{1}{2}})^4 = p^2$. Substitute $-\frac{1}{3}$ for $q$; $p^2 = (-\frac{1}{3})^{-2}$. A fraction with a negative exponent can be rewritten as the reciprocal of the fraction with a positive exponent; $(-\frac{1}{3})^{-2} = (-3)^2 = 9$; $p^2 = 9$, and $p = -3$ or 3.

**13. c.** Substitute $\frac{1}{3}$ for $a$ and 9 for $b$; $(\frac{1}{3}\sqrt{9}) = (\frac{1}{3})(3) = 1$; 1 is raised to the power $-3$, but the value of the exponent does not matter; 1 raised to any power is 1.

**14. b.** If $y = -x$, then $y = -2$. Substitute 2 for $x$ and $-2$ for $y$: $(((2)(-2))^{-2})^2 = ((-4)^{-2})^2 = (\frac{1}{16})^2 = \frac{1}{256}$.

**15. c.** First, cross multiply: $g(g\sqrt{108}) = \sqrt{3}$, $g^2\sqrt{108} = \sqrt{3}$. Divide both sides of the equation by $\sqrt{108}$: $g^2\sqrt{108} = \sqrt{3}$, $g^2 = \frac{\sqrt{3}}{\sqrt{108}}$, $g^2 = \frac{1}{\sqrt{36}}$, $g^2 = \frac{1}{6}$. Take the square root of both sides of the equation to find the value of $g$: $g^2 = \frac{1}{6}$, $g = \sqrt{\frac{1}{6}} = \frac{\sqrt{1}}{\sqrt{6}} = \sqrt{\frac{1}{6}}$. Simplify the fraction by multiplying it by $\frac{\sqrt{6}}{\sqrt{6}}$: $(\sqrt{\frac{1}{6}})(\frac{\sqrt{6}}{\sqrt{6}}) = \frac{\sqrt{6}}{\sqrt{6}}$.

**16. e.** First, substitute 2 for $c$. $(2\sqrt{d})^2 = 48$. $(2\sqrt{d})(2\sqrt{d}) = 4d$. $4d = 48$. Divide both sides of the equation by 4: $\frac{4d}{4} = \frac{48}{4}$, $d = 12$.

**17. d.** $\sqrt{n} = n^{\frac{1}{2}}$. The $n$ term in the denominator has a negative exponent. It can be placed in the numerator with a positive exponent, since $(\frac{1}{n^{-\frac{1}{2}}}) = n^{\frac{1}{2}}$. The numerator of the fraction is now $(n^{\frac{1}{2}})(n^{\frac{1}{2}})$ and the denominator of the fraction is 1. To multiply terms with like bases, keep the base and add the exponents: $(n^{\frac{1}{2}})(n^{\frac{1}{2}}) = n$. Therefore, $nm = 5$, and $m = \frac{5}{n}$.

**18. d.** First, multiply the first two terms. Multiply the coefficients of the terms (1 and 2) and add the exponents. Since $-y + y = 0$, $(x^{-y})(2x^y) = 2$, and $(x^{-y})(2x^y)(3y^x) = 2(3y^x) = 6y^x$. Substitute 2 for $x$ and $-2$ for $y$: $6(-2)^2 = 6(4) = 24$.

**19. e.** Substitute 20 for $n$: $\frac{\sqrt{20+5}}{\sqrt{20}}\left(\frac{20}{2}\sqrt{5}\right) = \frac{\sqrt{25}}{\sqrt{20}}$ $(10\sqrt{5}) = \frac{5}{2\sqrt{5}}(10\sqrt{5})$. Cancel the $\sqrt{5}$ terms and multiply the fraction by 10: $\frac{5}{2\sqrt{5}}(10\sqrt{5}) = \frac{5(10)}{2} = \frac{50}{2} = 25$.

**20. d.** If $a^2 = b = 4$, then $a = 2$. Substitute 2 for $a$ and 4 for $b$: $\left(\frac{4\sqrt{4}}{(2)4}\right)^2 = \left(\frac{(4)(2)}{16}\right)^2 = \left(\frac{8}{16}\right)^2 = \left(\frac{1}{2}\right)^2 = \frac{1}{4}$.

## ► Chapter 6

**1. c.** The fourth term in the sequence is 74. You are looking for the ninth term, which is 5 terms after the fourth term. Since each term is nine less than the previous term, the ninth term will be $5(9) = 45$ less than 74; $74 - 45 = 29$. Since the number of terms is reasonable, you can check your answer by repeatedly subtracting 9; $74 - 9 = 65$, $65 - 9 = 56$, $56 - 9 = 47$, $47 - 9 = 38$, $38 - 9 = 29$.

**2. d.** The fourth term in the sequence is $10\frac{1}{2}$. You are looking for the eighth term, which is four terms after the fourth term. Since each term is $\frac{3}{2}$ more than the previous term, the eighth term will be $4(\frac{3}{2}) = 6$ more than $10\frac{1}{2}$; $10\frac{1}{2} + 6 = 16\frac{1}{2}$. Since the number of terms is reasonable, you can check your answer by repeatedly adding $\frac{3}{2}$; $10\frac{1}{2} + \frac{3}{2} = 12$, $12 + \frac{3}{2} = 13\frac{1}{2}$, $13\frac{1}{2} + \frac{3}{2} = 15$, $15 + \frac{3}{2} = 16\frac{1}{2}$.

**3. e.** The term that precedes $x$ is 5. Therefore, the value of $x$ is $5 - 7 = -2$, and the value of $y$ is $-2 - 7 = -9$. Therefore, $x - y = -2 - (-9) = -2 + 9 = 7$.

**4. c.** The term that follows $z$ is 7. Since each term is 6 more than the previous term, $z$ must be 6 less than 7. Therefore, $z = 7 - 6 = 1$. In the same way, $y$ is 6 less than $z$ and $x$ is 6 less than $y$; $y = 1 - 6 = -5$ and $x = -5 - 6 = -11$. The sum of $x + z$ is equal to $-11 + 1 = -10$.

**5. e.** The first term in the sequence is 2. The next term in the sequence, $a$, is $\frac{1}{3}$ more than 2: $2\frac{1}{3}$. $b$ is $\frac{1}{3}$ more than $a$, $2\frac{2}{3}$. $c$ is $\frac{1}{3}$ more than 3: $3\frac{1}{3}$. $d$ is $\frac{1}{3}$ more than $c$, $3\frac{2}{3}$. Add the values of $a$, $b$, $c$, and $d$: $2\frac{1}{3} + 2\frac{2}{3} + 3\frac{1}{3} + 3\frac{2}{3} = 12$.

**6. b.** The fourth term in the sequence is $-24$. You are looking for the seventh term, which is three terms after the fourth term. You must multiply by $-2$ three times, so the seventh term will be $(-2)^3 = -8$ times $-24$; $(-24)(-8) = 192$. Since the number of terms is reasonable, you can check your answer by repeatedly multiplying by $-2$; $(-24)(-2) = 48$, $(48)(-2) = -96$, $(-96)(-2) = 192$.

**7. a.** The fourth term in the sequence is $\frac{16}{3}$. You are looking for the seventh term, which is three terms after the fourth term. You must multiply by $\frac{2}{3}$ three times, so the seventh term will be $(\frac{2}{3})^3 = \frac{8}{27}$ times $\frac{16}{3}$; $(\frac{8}{27})(\frac{16}{3}) = \frac{128}{81}$. Alternatively, every term in the sequence is 18 times $\frac{2}{3}$ raised to a power. The first term, 18, is $18 \times (\frac{2}{3})^0$. The second term, 12, is $18 \times (\frac{2}{3})^1$. The value of the exponent is one less than the position of the term in the sequence. The seventh term of the sequence is equal to $18 \times (\frac{2}{3})^6 = 18 \times (\frac{64}{729}) = 2 \times (\frac{64}{81}) = \frac{128}{81}$.

**8. a.** Every term in the sequence is 5 raised to a power. The first term, $\frac{1}{125}$, is $5^{-3}$. The second term, $\frac{1}{25}$, is $5^{-2}$. The value of the exponent is four less than the position of the term in the sequence. The 20th term of the sequence is equal to $5^{(20-4)} = 5^{16}$.

**9. b.** Since each term in the sequence is $-4$ times the previous term, $y$ is equal to $\frac{-64}{-4} = 16$, and $x = \frac{16}{-4} = -4$. Therefore, $xy = (16)(-4) = -64$.

**10. a.** Every term in the sequence is 3 raised to a power. The first term, 1, is $3^0$. The second term, 3, is $3^1$. The value of the exponent is one less than the position of the term in the sequence. The 100th term of the sequence is equal to $3^{(100-1)} = 3^{99}$ and the 101st term in the sequence is equal to $3^{(99+1)} = 3^{100}$. To multiply two terms with common bases, add the exponents of the terms: $(3^{99})(3^{100}) = 3^{199}$.

**11. b.** Since the rule of the sequence is each term is two less than three times the previous term, multiply the last term, $-53$, by $3$, then subtract $2$: $(-53)(3) = -159 - 2 = -161$.

**12. c.** Since the rule of the sequence is each term is nine more than $\frac{1}{3}$ the previous term, to find the value of $x$, multiply the last term, $36$, by $\frac{1}{3}$, then add $9$: $(36)(\frac{1}{3}) = 12$, $12 + 9 = 21$. In the same way, the value of $y$ is $21(\frac{1}{3}) + 9 = 7 + 9 = 16$. Therefore, the value of $y - x = 16 - 21 = -5$.

**13. e.** Since the rule of the sequence is each term is 20 less than five times the previous term, to find the value of $x$, add 20 to 0 and divide by 5: $\frac{(0 + 20)}{5} = \frac{20}{5} = 4$. In the same way, the value of $y$ is $\frac{(-120 + 20)}{5} = \frac{-100}{5} = -20$. Therefore, the value of $x + y = 4 + -20 = -16$.

**14. b.** Continue the sequence; 28.5 is the fourth term of the sequence. The fifth term is $(\frac{28.5}{2}) - 2 = 14.25 - 2 = 12.25$. The sixth term is $(\frac{12.25}{2}) - 2 = 6.125 - 2 = 4.125$, the seventh term is $(\frac{4.125}{2}) - 2 = 2.0625 - 2 = 0.0625$. Half of this number minus two will yield a negative value, so the eighth term of the sequence is the first term of the sequence that is a negative number.

**15. d.** Since the rule of the sequence is each term is 16 more than $-4$ times the previous term, to find the value of $y$, subtract 16 from $-80$ and divide by $-4$: $\frac{(-80 - 16)}{-4} = \frac{-96}{-4} = 24$. In the same way, the value of $x$ is $\frac{(24 - 16)}{-4} = \frac{8}{-4} = -2$. Therefore, the value of $x + y = -2 + 24 = 22$.

**16. d.** Since each term in the sequence below is equal to the sum of the two previous terms, $d = b + c$; $e = c + d$, since $c$ and $d$ are the two terms previous to $e$. If $e = c + d$, then, by subtracting $c$ from both sides of the equation, $d = e - c$. In the same way, $f = d + e$, the terms that precede it, and that equation can be rewritten as $d = f - e$; $d = b + c$, and $c = a + b$. Therefore, $d = b + (a + b)$, $d = a + 2b$. However, $d$ is not equal to $e - 2b$; $d = e - c$, and $c = a + b$, not $2b$, since $a$ is not equal to $b$.

## ▶ Chapter 7

**1. d.** Solve the first equation for $y$ in terms of $x$; $2x + y = 6$, $y = 6 - 2x$. Substitute this expression for $y$ in the second equation and solve for $x$:

$$\frac{6 - 2x}{2 + 4x} = 12$$
$$3 - x + 4x = 12$$
$$3x + 3 = 12$$
$$3x = 9$$
$$x = 3$$

**2. c.** Add the two equations together. The $b$ terms will drop out, and you can solve for $a$:

$$5a + 3b = -2$$
$$+\ 5a - 3b = -38$$
$$10a = -40$$
$$a = -4$$

Substitute $-4$ for $a$ in the first equation and solve for $b$:

$$5(-4) + 3b = -2$$
$$-20 + 3b = -2$$
$$3b = 18$$
$$b = 6$$

**3. a.** Solve the first equation for $x$ in terms of $y$; $xy = 32$, $x = \frac{32}{y}$. Substitute this expression for $x$ in the second equation and solve for $y$:

$$2x - y = 0$$
$$2(\frac{32}{y}) - y = 0$$
$$\frac{64}{y} - y = 0$$
$$\frac{64}{y} = y$$
$$y^2 = 64$$
$$y = -8, y = 8$$

**4. b.** In the first equation, multiply the $(x + 4)$ term by 3: $3(x + 4) = 3x + 12$. Then, subtract 12 from both sides of the equation, and the first equation becomes $3x - 2y = -7$. Add the two equations together. The $y$ terms will drop out, and you can solve for $a$:

$$3x - 2y = -7$$
$$+\ 2y - 4x = 8$$
$$-x = 1$$
$$x = -1$$

**5. e.** Solve the second equation for $a$ in terms of $b$; $b + a = 13$, $a = 13 - b$. Substitute this expression for $a$ in the first equation and solve for $b$:

$$-7a + \frac{b}{4} = 25$$
$$-7(13 - b) + \frac{b}{4} = 25$$
$$7b - 91 + \frac{b}{4} = 25$$
$$\frac{29b}{4} = 116$$
$$29b = 464$$
$$b = 16$$

**6. c.** Solve the second equation for $x$ in terms of $y$; $\frac{4y}{x} = 1$, $x = 4y$. Substitute this expression for $x$ in the first equation and solve for $y$:

$$3x + 7y = 19$$
$$3(4y) + 7y = 19$$
$$12y + 7y = 19$$
$$19y = 19$$
$$y = 1$$

**7. b.** In the first equation, multiply the $(m + n)$ term by 2 and add $m$: $2(m + n) + m = 2m + 2n + m = 3m + 2n$. Subtract the second equation from the first equation. The $m$ terms will drop out, and you can solve for $n$:

$$3m + 2n = 9$$
$$\underline{-3m - 3n = -24}$$
$$5n = -15$$
$$n = -3$$

**8. d.** In the first equation, multiply the $(b + 4)$ term by $-2$: $-2(b + 4) = -2b - 8$. Add 8 to both sides of the equation, and the first equation becomes $9a - 2b = 38$. Multiply the second equation by 2 and subtract it from the first equation. The $a$ terms will drop out, and you can solve for $b$:

$$2(4.5a - 3b = 3) = 9a - 6b = 6$$
$$9a - 2b = 38$$
$$\underline{-(9a - 6b = 6)}$$
$$4b = 32$$
$$b = 8$$

**9. b.** Solve the second equation for $q$ in terms of $p$; $4p - 2q = -14$, $-2q = -4p - 14$, $q = 2p + 7$. Substitute this expression for $q$ in the first equation and solve for $p$:

$$4pq - 6 = 10$$
$$4p(2p + 7) - 6 = 10$$

$$8p^2 + 28p - 6 = 10$$
$$8p^2 + 28p - 16 = 0$$
$$2p^2 + 7p - 4 = 0$$
$$(2p - 1)(p + 4) = 0$$
$$2p - 1 = 0, 2p = 1, p = \frac{1}{2}$$
$$p + 4 = 0, p = -4$$

**10. a.** Solve the second equation for $b$ in terms of $a$; $b + 2a = -4$, $b = -2a - 4$. Substitute this expression for $b$ in the first equation and solve for $a$:

$$7(2a + 3(-2a - 4)) = 56$$
$$7(2a + -6a - 12) = 56$$
$$7(-4a - 12) = 56$$
$$-28a - 84 = 56$$
$$-28a = 140$$
$$a = -5$$

**11. c.** Multiply the first equation by 8 and add it to the second equation. The $x$ terms will drop out, and you can solve for $y$:

$$8(\tfrac{1}{2}x + 6y = 7) = 4x + 48y = 56$$
$$4x + 48y = 56$$
$$\underline{+ -4x - 15y = 10}$$
$$33y = 66$$
$$y = 2$$

**12. a.** Solve the second equation for $n$ in terms of $m$; $m - n = 0$, $n = m$. Substitute this expression for $n$ in the first equation and solve for $m$:

$$m(n + 1) = 2$$
$$m(m + 1) = 2$$
$$m^2 + m = 2$$
$$m^2 + m - 2 = 0$$
$$(m + 2)(m - 1) = 0$$
$$m + 2 = 0, m = -2$$
$$m - 1 = 0, m = 1$$

**13. b.** Solve the second equation for $c$ in terms of $d$; $c - 6d = 0$, $c = 6d$. Substitute this expression for $c$ in the first equation and solve for $d$:

$$\frac{c - d}{5} - 2 = 0$$
$$\frac{6d - d}{5} - 2 = 0$$
$$\frac{5d}{5} - 2 = 0$$
$$d - 2 = 0$$
$$d = 2$$

Substitute the value of $d$ into the second equation and solve for $c$:

$$c - 6(2) = 0$$
$$c - 12 = 0$$
$$c = 12$$

Since $c = 12$ and $d = 2$, the value of $\frac{c}{d} = \frac{12}{2} = 6$.

**14. d.** Divide the second equation by 2 and add it to the first equation. The $b$ terms will drop out, and you can solve for $a$:

$$\frac{(6a - 12b = -6)}{2} = 3a - 6b = -3$$
$$4a + 6b = 24$$
$$\underline{+\ 3a - 6b = -3}$$
$$7a = 21$$
$$a = 3$$

Substitute the value of $a$ into the first equation and solve for $b$:

$$4(3) + 6b = 24$$
$$12 + 6b = 24$$
$$6b = 12$$
$$b = 2$$

Since $a = 3$ and $b = 2$, the value of $a + b = 3 + 2 = 5$.

**15. e.** Multiply the first equation by $-6$ and add it to the second equation. The $x$ terms will drop out, and you can solve for $y$:

$$-6(\tfrac{x}{3} - 2y = 14) = -2x + 12y = -84$$
$$-2x + 12y = -84$$
$$\underline{+\ 2x + 6y = -6}$$
$$18y = -90$$
$$y = -5$$

Substitute the value of $y$ into the second equation and solve for $x$:

$$2x + 6(-5) = -6$$
$$2x - 30 = -6$$
$$2x = 24$$
$$x = 12$$

Since $x = 12$ and $y = -5$, the value of $x - y = 12 - (-5) = 12 + 5 = 17$.

**16. a.** Solve the second equation for $y$ in terms of $x$; $-x - y = -6$, $-y = x - 6$, $y = -x + 6$. Substitute this expression for $y$ in the first equation and solve for $x$:

$$-5x + 2(-x + 6) = -51$$
$$-5x - 2x + 12 = -51$$
$$-7x + 12 = -51$$
$$-7x = -63$$
$$x = 9$$

Substitute the value of $x$ into the second equation and solve for $y$:

$$-9 - y = -6$$
$$-y = 3$$
$$y = -3$$

Since $x = 9$ and $y = -3$, the value of $xy = (9)(-3) = -27$.

**17. c.** First, multiply $(n + 2)$ by $-6$: $-6(n + 2) = -6n - 12$. Then, add 12 to both sides of the equation. The first equation becomes $m - 6n = 4$. Add the two equations. The $n$ terms will drop out, and you can solve for $m$:

$$m - 6n = 4$$
$$\underline{+\ 6n + m = 16}$$
$$2m = 20$$
$$m = 10$$

Substitute the value of $m$ into the second equation and solve for $n$:

$$6n + 10 = 16$$
$$6n = 6$$
$$n = 1$$

Since $m = 1$ and $n = 10$, the value of $\frac{n}{m} = \frac{1}{10}$.

**18. d.** First, simplify the first equation by subtracting 4 from both sides. The first equation becomes $3x = -5y + 4$. Then, multiply the equation by $-3$ and add it to the second equation. The $x$ terms will drop out, and you can solve for $y$:

$$-3(3x = -5y + 4) = -9x = 15y - 12$$
$$-9x = 15y - 12$$
$$\underline{+\ 9x + 11y = -8}$$
$$11y = 15y - 20$$
$$-4y = -20$$
$$y = 5$$

Substitute the value of $y$ into the second equation and solve for $x$:

$$9x + 11(5) = -8$$

$$9x + 55 = -8$$
$$9x = -63$$
$$x = -7$$

Since $y = 5$ and $x = -7$, the value of $y - x = 5 - (-7) = 5 + 7 = 12$.

**19. b.** First, simplify the first equation by multiplying ($a + 3$) by $\frac{1}{2}$. The first equation becomes $\frac{1}{2}a + \frac{3}{2} - b = -6$. Subtract $\frac{3}{2}$ from both sides, and the equation becomes $\frac{1}{2}a - b = -\frac{15}{2}$. Then, multiply the equation by $-6$ and add it to the second equation. The $a$ terms will drop out, and you can solve for $b$:

$$-6(\tfrac{1}{2}a - b = -\tfrac{15}{2}) = -3a + 6b = 45$$
$$-3a + 6b = 45$$
$$\underline{+\ 3a - 2b = -5}$$
$$4b = 40$$
$$b = 10$$

Substitute the value of $b$ into the second equation and solve for $a$:

$$3a - 2(10) = -5$$
$$3a - 20 = -5$$
$$3a = 15$$
$$a = 5$$

Since $a = 5$ and $b = 10$, the value of $a + b = 5 + 10 = 15$.

**20. e.** Solve the second equation for $b$ in terms of $a$; $b - a = 1$, $b = a + 1$. Substitute this expression for $b$ in the first equation and solve for $a$:

$$10(a + 1) - 9a = 6$$
$$10a + 10 - 9a = 6$$
$$a + 10 = 6$$
$$a = -4$$

Substitute the value of $a$ into the second equation and solve for $b$:

$$b - (-4) = 1$$
$$b + 4 = 1$$
$$b = -3$$

Since $a = -4$ and $b = -3$, the value of $ab = (-4)(-3) = 12$.

**21. b.** Solve the second equation for $y$ in terms of $x$; $2x - y = 9$, $-y = -2x + 9$, $y = 2x - 9$. Substitute this expression for $y$ in the first equation and solve for $x$:

$$\frac{x + 2x - 9}{3} = 8$$
$$\frac{3x - 9}{3} = 8$$
$$x - 3 = 8$$
$$x = 11$$

Substitute the value of $x$ into the second equation and solve for $y$:

$$2(11) - y = 9$$
$$22 - y = 9$$
$$-y = -13$$
$$y = 13$$

Since $x = 11$ and $y = 13$, the value of $x - y = 11 - 13 = -2$.

**22. b.** First, simplify the second equation by subtracting 9 from both sides of the equation. The second equation becomes $-2x - 6 = y$. Then, multiply the equation by 2 and add it to the first equation. The $x$ terms will drop out, and you can solve for $y$:

$$2(-2x - 6 = y) = -4x - 12 = 2y$$
$$-4x - 12 = 2y$$
$$\underline{+\ 4x + 6 = -3y}$$
$$-6 = -y$$
$$y = 6$$

Substitute the value of $y$ into the first equation and solve for $x$:

$$4x + 6 = -3(6)$$
$$4x + 6 = -18$$
$$4x = -24$$
$$x = -6$$

Since $x = -6$ and $y = 6$, the value of $\frac{x}{y} = -\frac{6}{6} = -1$.

**23. c.** Multiply the second equation by 3 and add it to the first equation. The $p$ terms will drop out, and you can solve for $q$:

$$3(-5p + 2q = 24) = -15p + 6q = 72$$
$$-15p + 6q = 72$$
$$\underline{+\ 8q + 15p = 26}$$
$$14q = 98$$
$$q = 7$$

Substitute the value of $q$ into the second equation and solve for $p$:

$$-5p + 2(7) = 24$$
$$-5p + 14 = 24$$
$$-5p = 10$$
$$p = -2$$

Since $p = -2$ and $q = 7$, the value of $(p + q)^2 = (-2 + 7)^2 = 5^2 = 25$.

**24. e.** First, simplify the first equation by multiplying $(x - 1)$ by 9: $9(x - 1) = 9x - 9$. Then, add 9 and $4y$ to both sides of the equation. The first equation becomes $9x + 4y = 11$. Then, multiply the second equation by $-2$ and add it to the first equation. The $y$ terms will drop out, and you can solve for $x$:

$$-2(2y + 7x = 3) = -4y - 14x = -6$$
$$-4y - 14x = -6$$
$$\underline{+\ 9x + 4y = 11}$$
$$-5x = 5$$
$$x = -1$$

Substitute the value of $x$ into the second equation and solve for $y$:

$$2y + 7(-1) = 3$$
$$2y - 7 = 3$$
$$2y = 10$$
$$y = 5$$

Since $y = 5$ and $x = -1$, the value of $(y - x)^2 = (5 - (-1))^2 = 6^2 = 36$.

**25. d.** Solve the first equation for $a$ in terms of $b$ by multiplying both sides of the equation by 2. $a = 2b + 2$. Substitute this expression for $a$ in the second equation:

$$3(2b + 2 - b) = -21$$
$$3(b + 2) = -21$$
$$3b + 6 = -21$$
$$3b = -27$$
$$b = -9$$

Substitute the value of $b$ into the first equation and solve for $a$:

$$\frac{a}{2} = -9 + 1$$
$$\frac{a}{2} = -8$$
$$a = -16$$

Since $a = -16$ and $b = -9$, the value of $\sqrt{\frac{a}{b}}$
$$= \sqrt{\frac{-16}{-9}} = \sqrt{\frac{16}{9}} = \frac{4}{3}.$$

## ▶ Chapter 8

**1. c.** Draw a horizontal line across the coordinate plane where $f(x) = 2$. In other words, graph the line $y = 2$. This line touches the graph of $f(x)$ in 3 places. Therefore, there are 3 values for which $f(x) = 2$.

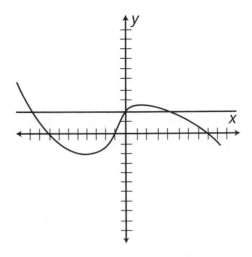

**2. b.** Draw a horizontal line across the coordinate plane where $f(x) = 3$. This line touches the graph of $f(x)$ in 1 place. Therefore, there is 1 value for which $f(x) = 3$.

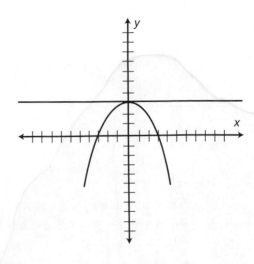

**3. d.** $f(x) = 0$ every time the graph touches the $x$-axis, since the $x$-axis is the graph of the line $f(x) = 0$. The graph of $f(x)$ touches the $x$-axis in 5 places. Therefore, there are 5 values for which $f(x) = 0$.

**4. e.** Draw a horizontal line across the coordinate plane where $f(x) = -5$. This line touches the graph of $f(x)$ in 8 places. Therefore, there are 8 values for which $f(x) = -5$.

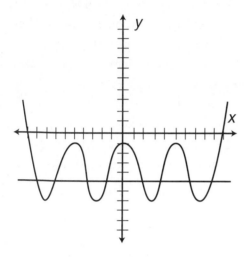

**5. a.** Draw a horizontal line across the coordinate plane where $f(x) = -2$. This line does not touch the graph of $f(x)$ at all. Therefore, there are 0 values for which $f(x) = -2$.

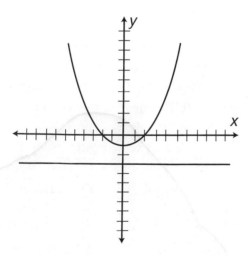

**6. d.** Begin with the innermost function: find $g(-3)$ by substituting $-3$ for $x$ in the function $g(x)$:

$g(-3) = (-3)^2 = 9$. Then, substitute the result of that function for $x$ in $f(x)$; $f(9) = 2(9) - 1 = 18 - 1 = 17$.

**7. b.** Begin with the innermost function: Find $f(-2)$ by substituting $-2$ for $x$ in the function $f(x)$: $f(-2) = 3(-2) + 2 = -6 + 2 = -4$. Then, substitute the result of that function for $x$ in $g(x)$; $g(-4) = 2(-4) - 3 = -8 - 3 = -11$.

**8. e.** Begin with the innermost function: Find $f(3)$ by substituting 3 for $x$ in the function $f(x)$: $f(3) = 2(3) + 1 = 6 + 1 = 7$. Next, substitute the result of that function for $x$ in $g(x)$; $g(7) = 7 - 2 = 5$. Finally, substitute 5 for $x$ in $f(x)$: $f(5) = 2(5) + 1 = 10 + 1 = 11$.

**9. c.** Begin with the innermost function. You are given the value of $f(x)$: $f(x) = 6x + 4$. Substitute this expression for $x$ in the equation $g(x)$:

$g(x) = x^2 - 1$
$g(6x + 4) = (6x + 4)^2 - 1$
$g(6x + 4) = 36x^2 + 24x + 24x + 16 - 1$
$g(6x + 4) = 36x^2 + 48x + 15$
Therefore, $g(f(x)) = 36x^2 + 48x + 15$

**10. b.** Begin with the innermost function. You are given the value of $g(x)$: $g(x) = 2\sqrt{x}$. Substitute this expression for $x$ in the equation $f(x)$:

$f(x) = 4 - 2x^2$
$f(2\sqrt{x}) = 4 - 2(2\sqrt{x})^2$
$f(2\sqrt{x}) = 4 - 2(4x)$
$f(2\sqrt{x}) = 4 - 8x$
Therefore, $f(g(x)) = 4 - 8x$

**11. c.** The definition of the function states that you must multiply by 3 the term before the @ symbol, then subtract from that product the term after the @ symbol. Begin with the function in parentheses, $w@z$, which is given as $3w - z$. Replace the expression $w@z$ with $3w - z$, and the problem becomes $(3w - z)@z$. Again, multiply the term before the @ symbol by 3, and subtract from it the term after the @ symbol. Multiply $3w - z$ by 3 and subtract $z$: $(3w - z)@z = 3(3w - z) - z = 9w - 3z - z = 9w - 4z$.

**12. b.** The definition of the function states that the term before the & symbol should be divided by the term after the & symbol, and that quotient should be added to the product of the two terms. Therefore, the value of $q\&p = \frac{q}{p} + qp$. Substitute 4 for $p$ and $-2$ for $q$ in this definition: $-\frac{2}{4} + (-2)(4) = -\frac{1}{2} - 8 = -8.5$.

**13. a.** The definition of the function states that the term after the % symbol should be raised to the power of the term before the % symbol. Replace the expression $j\%k$ with its given value, $k^j$, and the problem becomes $k\%(k^j)$. The term after the % symbol, $k^j$, should be raised to the power of the term before the % symbol, $k$: $(k^j)^k = k^{jk}$. Remember, when an exponent is raised to an exponent, multiply the exponents.

**14. e.** The definition of the function states that the term before the ? symbol should be subtracted from the term after the ? symbol, and that difference should be divided by the sum of the two terms. Begin with the innermost function: $a?b$, which is given as $\frac{b-a}{a+b}$. Substitute the values of $a$ and $b$: $\frac{b-a}{a+b} = \frac{-5-6}{6+(-5)} = \frac{-11}{1} = -11$. Now evaluate $a?-11$. Remember, the term before the ? symbol should be subtracted from the term after the ? symbol, and that difference should be divided by the sum of the two terms. Substitute the value of $a$ again: $\frac{-11-6}{6+(-11)} = \frac{-17}{-5} = \frac{17}{5}$.

**15. d.** The definition of the function states that the term before the ^ symbol should be squared and the term after the ^ symbol should be subtracted from that square. Begin with the innermost function: $y^\wedge y$: Square the second term, $y$, and subtract the first term, $y$: $y^2 - y$. Now evaluate $y^\wedge(y^2 - y)$. Again, square the second term, $y^2 - y$, and subtract the first term, $y$: $(y^2 - y)^2 - y = y^4 - y^3 - y^3 + y^2 - y = y^4 - 2y^3 + y^2 - y$.

**16. b.** The domain of a function is the set of all possible inputs to the function. All real numbers can be substituted for $x$ in the function $f(x) = \frac{1}{(x^2 - 9)}$ excluding those that make the fraction undefined. Set the denominator equal to 0 to find the

$x$ values that make the fraction undefined. These values are not in the domain of the function; $x^2 - 9 = 0$, $(x - 3)(x + 3) = 0$, $x - 3 = 0$, $x = 3$; $x + 3 = 0$, $x = -3$. The domain of $f(x)$ is all real numbers excluding 3 and $-3$.

**17. e.** The range of a function is the set of all possible outputs of the function. All real numbers can be substituted for $x$ in the function $f(x) = x^2 - 4$, so the domain of the function is all real numbers. Since the $x$ term is squared, the smallest value that this term can equal is 0 (when $x = 0$). Therefore, the smallest value that $f(x)$ can have is when $x = 0$. When $x = 0$, $f(x) = 0^2 - 4 = -4$. The range of $f(x)$ is all real numbers greater than or equal to $-4$.

**18. a.** The domain of the function is all real numbers; any real number can be substituted for $x$. There are no $x$ values that can make the function undefined. The range of a function is the set of all possible outputs of the function. Since the $x$ term is squared, then made negative, the largest value that this term can equal is 0 (when $x = 0$). Every other $x$ value will result in a negative value for $f(x)$. The range of $f(x)$ is all real numbers less than or equal to 0.

**19. c.** The square root of a negative value is imaginary, so the value of $4x - 1$ must be greater than or equal to 0. $4x - 1 \geq 0$, $4x \geq 1$, $x \geq \frac{1}{4}$. The domain of $f(x)$ is all real numbers greater than or equal to $\frac{1}{4}$. Since $x$ must be greater than or equal to $\frac{1}{4}$, the smallest value of $f(x)$ is the square root of 0, which is 0. The range of the function is all real numbers greater than or equal to 0.

**20. c.** The square root of a negative value is imaginary, so the value of $x - 5$ cannot be negative. Since the square root is the denominator of a fraction, it cannot be equal to 0 either. The value of $\sqrt{x - 5}$ must be greater than 0; $\sqrt{x - 5} > 0$, $x - 5 > 0$, $x > 5$. The domain of $f(x)$ is all real numbers greater than 5. Since the denominator of the fraction must always be positive, and the numerator of the fraction is $-1$, the value of $f(x)$ will always be negative. No value of $x$ will make $f(x) = 0$, so the range of $f(x)$ is all real numbers less than 0.

**21. a.** The graph of the equation in diagram $A$ is not a function. A function is an equation in which each unique input yields no more than one output. The equation in diagram $A$ fails the vertical line test for all $x$ values where $-2 < x < 2$. For each of these $x$ values (inputs), there are two $y$ values (outputs).

**22. d.** The range of a function is the set of possible outputs of the function. In each of the five equations, the set of possible $y$ values that can be generated for the equation is the range of the equation. Find the coordinate planes that show a graph that extends below the $x$-axis. These equations have negative $y$ values, which means that the range of the equation contains negative values. The graphs of the equations in diagrams $A$, $B$, and $D$ extend below the $x$-axis. However, the graph of the equation in diagram $A$ is not a function. It fails the vertical line test for all $x$ values where $-2 < x < 2$. The equations graphed in diagrams $B$ and $D$ are functions whose ranges contain negative values.

**23. e.** The equation of the graph in diagram $B$ is $y = |x| - 3$. Any real number can be substituted into this equation. There are no $x$ values that will generate an undefined or imaginary $y$ value. The equation of the graph in diagram $E$ is $y = (x-3)^2 + 1$. With this equation as well, any real number can be substituted for $x$—there are no $x$ values that will generate an undefined or imaginary $y$ value. The equation of the graph in diagram $D$ is $y = \frac{1}{x}$. If $x = 0$, this function will be undefined. Therefore, the domain of this function is all real numbers excluding 0. Only the functions in diagrams $B$ and $E$ have a domain of all real numbers with no exclusions.

**24. b.** The equation of the graph in diagram $C$ is $y = \sqrt{x}$. Since the square root of a negative number is imaginary, the domain of this equation is all

real number[s]
square roots of
equal to 0 are also r[eal]
than or equal to 0. Th[e]
equation $y = \sqrt{x}$ is all real [numbers greater than or equal to]
0, and the domain and range o[f]
the same. The equation of the gra[ph in diagram]
$D$ is $y = \frac{1}{x}$. If $x = 0$, this function wi[ll be unde-]
fined. Therefore, the domain of this fun[ction is]
all real numbers excluding 0. One divided [by any]
real number (excluding 0) will yield real num[-]
bers, excluding 0. Therefore, the range of the equation $y = \frac{1}{x}$ is all real numbers excluding 0, and the domain and range of the equation are the same. The equation of the graph in diagram $B$ is $y = |x| - 3$. Any real number can be substituted into this equation. There are no $x$ values that will generate an undefined or imaginary $y$ value. However, it is impossible to generate a $y$ value that is less than $-3$. Any $x$ value greater than or less than 3 will generate a $y$ value that is greater than $-3$. Therefore, the range of the equation $y = |x| - 3$ is all real numbers greater than or equal to $-3$. The domain and range of $y = |x| - 3$ are not the same. The equation of the graph in diagram $E$ is $y = (x-3)^2 + 1$. With this equation as well, any real number can be substituted for $x$—there are no $x$ values that will generate an undefined or imaginary $y$ value. However, it is impossible to generate a $y$ value that is less than 1. Any $x$ value greater than or less than 3 will generate a $y$ value that is greater than 1. Therefore, the range of the equation $y = (x-3)^2 + 1$ is all real numbers greater than or equal to 1. The domain and range of $y = (x-3)^2 + 1$ are not the same.

**25. e.** The graph of the equation in diagram $A$ is not a function. It fails the vertical line test for all $x$ values where $-2 < x < 2$. The equation of the graph in diagram $B$ is $y = |x| - 3$. Any $x$ value greater than or less than 3 will generate a $y$ value that is greater than $-3$; no values less than

greater than or equal to 0. The
real numbers greater than or
real numbers that are greater
erefore, the range of the
numbers greater than
f the equation are
ph in diagram
ll be unde-
ction is
by a

than 3 will generate a *y* value that is greater than 1; no values less than 1 can be generated by this equation. Therefore, the range of the equation $y = (x-3)^2 + 1$ is all real numbers greater than or equal to 1. Of the four equations that are functions, the equation $y = (x-3)^2 + 1$ (*E*), has the smallest range (fewest elements), since the set of real numbers that are greater than or equal to 1 is smaller than the set of all real numbers greater than or equal to −3 (*B*), smaller than the set of all real numbers greater than or equal to 0 (*C*), and smaller than the set of all real numbers excluding 0 (*D*).

## ▶ Chapter 9

**1. d.** Angles 2, 4, 6, and 7 are alternating (vertical) angles. Therefore, their measures are equal. Angles 7 and 8 are supplementary. Therefore, angles 2 and 8 are also supplementary; $12x + 10 + 7x - 1 = 180$, $19x + 9 = 180$, $19x = 171$, $x = 9$. Since $x = 9$, the measure of angle 2 is $12(9) + 10 = 108 + 10 = 118°$.

**2. e.** Angles 5 and 6 are supplementary. Therefore, the sum of their measures is 180°. If the measure of angle 6 is $x$, then the measure of angle 5 is $5x$; $5x + x = 180$, $6x = 180$, $x = 30$. The measure of angle 6 is 30°, and the measure of angle 5 is $5(30) = 150°$.

**3. c.** Angles 4 and 7 are alternating angles. Therefore, their measures are equal; $6x + 20 = 10x - 40$, $4x = 60$, $x = 15$. Since $x = 15$, the measure of angles 4 and 7 is $6(15) + 20 = 90 + 20 = 110$. Notice that replacing $x$ with 15 in the measure of angle 7 also yields 110: $10(15) - 40 = 150 - 40 = 110$. Since angles 6 and 7 are vertical angles, the measure of angle 6 is also 110°.

**4. a.** If angle 3 measures 90°, then angles 1, 6, and 7 must also measure 90°, since they are alternating angles. Angles 3 and 4 are supplementary, since these angles form a line. Therefore, the measure of angle 4 is equal to $180 - 90 = 90°$. Angles 3 and 4 are congruent and supplementary. Since angles 2, 4, 5, and 8 are alternating angles, they are all congruent to each other. Every numbered angle measures 90°. Therefore, every numbered angle is congruent and supplementary to every other numbered angle. Angles 5 and 7 are in fact adjacent, since they share a common vertex and a common ray. However, angles 1 and 2 are not complementary—their measures add to 180°, not 90°.

**5. b.** Angles 2 and 6 are alternating angles. Therefore, their measures are equal; $8x + 10 = x^2 - 38$, $x^2 - 8x - 48 = 0$. Factor $x^2 - 8x - 48$ and set each factor equal to 0; $x^2 - 8x - 48 = (x + 4)(x - 12)$, $x + 4 = 0$, $x = -4$; $x - 12 = 0$, $x = 12$. An angle cannot have a negative measure, so the −4 value of $x$ must be discarded. If $x = 12$, then the measure of angle 2 is $8(12) + 10 = 106°$. Notice that replacing $x$ with 12 in the measure of angle 6 also yields 106: $(12)^2 - 38 = 144 - 38 = 106°$. Since angles 6 and 8 are supplementary, the measure of angle 8 is equal to $180 - 106 = 74°$.

**6. a.** Perpendicular lines cross at right angles. Therefore, angle *AOC* is 90°. Since angles 1 and 2 combine to form angle *AOC*, the sum of the measures of angles 1 and 2 must be 90°. Therefore, they are complementary angles.

**7. d.** Since line *AE* is perpendicular to ray *OC*, angles *AOC* and *EOC* are both 90°. Since angles 1 and 2 combine to form angle *AOC* and angles 3 and

4 combine to form angle *EOC*, these sums must both equal 90°. Therefore, angle 1 + angle 2 = angle 3 + angle 4. Angles 1, 2, 3, and 7 form a line, as to angles 4, 5, and 6. Therefore, the measures of angles 1, 2, 3, and 7 add to 180°, as do the sum of the measures of angles 4, 5, and 6. In the same way, angles 2, 3, 4, and 5 form a line, so the sum of the measures of those angles is also 180°. Angles *GOF* and *BOD* are vertical angles. Therefore, their measures are equal. Since angle 6 is angle *GOF* and angles 2 and 3 combine to form angle *BOD*, the measure of angle 6 is equal to the sum of the measures of angles 2 and 3. However, the sum of angle 1 and angle 7 is not equal to the sum of angle 2 and angle 3. In fact, the sum of angles 2 and 3 is supplementary to the sum of angles 7 and 1, since angle 6 is supplementary to the sum of angles 7 and 1, and the sum of angles 2 and 3 is equal to the measure of angle 6.

**8. c.** Since line *AE* is perpendicular to ray *OC*, angle *EOC* is a right angle (measuring 90°). Angles 3 and 4 combine to form angle *EOC*; therefore, their sum is equal to 90°: $2x + 2 + 5x - 10 = 90$, $7x - 8 = 90$, $7x = 98$, $x = 14$. Angle 4 is equal to $5(14) - 10 = 70 - 10 = 60°$. Since angles 4 and 7 are vertical angles, their measures are equal, and angle 7, too, measures 60°.

**9. b.** Since line *AE* is perpendicular to ray *OC*, angle *EOC* is a right angle (measuring 90°). Angles 3 and 4 combine to form angle *EOC*; therefore, their sum is equal to 90°: $90 - 57 = 33$. Angle 3 measures 33°. Angle *AOC* is also a right angle, with angles 1 and 2 combining to form that angle. Therefore, the measure of angle 2 is equal to: $90 - 62 = 28°$. Angle 6 and angle *BOD* are vertical angles; their measures are equal. Since angles 2 and 3 combine to form angle *BOD*, the measure of *BOD*, and angle 6, is equal to: $33 + 28 = 61°$.

**10. b.** Since line *AE* is perpendicular to ray *OC*, angle *EOC* is a right angle (measuring 90°). Angles 3

and 4 combine to form angle *EOC*; therefore, their sum is equal to 90°: $5x + 3 + 15x + 7 = 90$, $20x + 10 = 90$, $20x = 80$, $x = 4$. Therefore, the measure of angle 4 is equal to: $15(4) + 7 = 67°$. Angles 4, 5, and 6 form a line; therefore, the sum of their measures is 180°. If *x* is the sum of angles 5 and 6, then $67 + x = 180$, and $x = 113°$.

**11. a.** Since angles *AIC* and *KIL* are vertical angles, their measures are equal.

**12. d.** Since line *IJ* is perpendicular to line *GH*, angle *IJL* is a right angle, which measures 90°. There are 180° in a triangle, so the measures of angles *JLI* and *JIL* must add to 90°: $8x - 4 + 5x + 3 = 90$, $13x - 1 = 90$, $13x = 91$, $x = 7$. Therefore, the measure of angle *JLI* is $8(7) - 4 = 52°$. Since angles *JLI* and *AIE* are alternating angles, their measures are equal. Angle *AIE* is also 52°.

**13. b.** Angles *GKI* and *EIC* are alternating angles; therefore, angle *EIC* also measures $15x - 4$. Since angles *EIC* and *CIF* form a line, their measures add to 180°; $15x - 4 + x^2 = 180$, $x^2 + 15x - 184 = 0$. Factor and solve for *x*: $(x + 23)(x - 8) = 0$, $x + 23 = 0$, $x = -23$. An angle cannot have a negative value, so disregard the negative value of *x*; $x - 8 = 0$, $x = 8$. Therefore, the measure of angle *EIC* is equal to: $15(8) - 4 = 120 - 4 = 116°$.

**14. a.** Since line *IJ* is perpendicular to line *EF*, angles *JIF* and *EIJ* are right angles, which measures 90° each. Angles *LIF* and *JIL* combine to form angle *JIF*; therefore, their measures add to 90°. If the measure of angle *JIL* is *x*, then the measure of angle *LIF* is $x + 6$, since its measure is 6 greater than the measure of angle *JIL*; $x + x + 6 = 90$, $2x + 6 = 90$, $2x = 84$, $x = 42$. The measure of angle *JIL* is 42° and the measure of angle *LIF* is $42 + 6 = 48°$. Since angles *LIF* and *EIK* are congruent, angle *EIK* is also 48°. Angles *EIK* and *KIJ* are complementary, so angle *KIJ* is $90 - 48 = 42°$.

**15. e.** Angles *JLB*, *ILH*, and *EIL* are alternating angles; therefore, their measures are equal. Since angles *EIL* and *LIF* form a line, these angles are supplementary and add to 180°. If angle *LIF* is *x*, then angle *JLB* is 3.5*x*, and *x* + 3.5*x* = 180, 4.5*x* = 180, *x* = 40. Therefore, the measure of angle *LIF* is 40° and the measure of angle *EIL* is 3.5(40) = 140°.

**16. e.** Since lines *EF* and *AB* are perpendicular, angle *AOE* measures 90°. Angle *AOE* is made up of angles 2 and 3. Angles 1, 2, 3, and 4 form a line. Since the measure of angles 2 and 3 add to 90°, and there are 180° in a line, the measures of angles 1 and 4 must add to 90°; 11*x* + 7*x* = 90, 18*x* = 90, *x* = 5. Therefore, the measure of angle 4 is 11(5) = 55°.

**17. e.** Angles 4 and 8 are vertical angles, angles 1 and 5 are vertical angles, and angles 2 and 6 are vertical angles. Since vertical angles have equal measures, the sum of angles 4, 5, and 6 is equal to the sum of angles 1, 2, and 8. Lines *AB* and *EF* are not known to be perpendicular. Test the other answer choices with different possible values. Angles 3 and 7 could measure 40°, while angles 1, 2, 5, and 6 measure 45°, and angles 4 and 8 measure 50°. Using these numbers, only the number sentence in choice **e** holds true.

**18. a.** Angles 2 and 6 are vertical angles and angles 3 and 7 are vertical angles. Therefore, angle 2 is equal to angle 6, angle 3 is equal to angle 7, and the sum of angles 2 and 3 is equal to the sum of angles 6 and 7: $x^2 = 10x$; $x^2 - 10x = 0$, $x(x - 10) = 0$. *x* cannot be 0, since no angles measure 0°; $x - 10 = 0$, $x = 10$, and the sum of angles 2 and 3 is $(10)^2 = 100$. Since angles 2, 3, 4, and 5 form a line, the sum of their measures is 180°. Therefore, the sum of angles 4 and 5 is equal to 180 − 100 = 80°.

**19. c.** Angles 2 and 6 are vertical angles; their measures are equal. Therefore, the measure of angle 2 also 5*x* − 3. Since lines *CD* and *GH* are perpendicular to each other, angle *COG* is a right angle, and angles 1 and 2 are complementary; 3*x* + 5 + 5*x* − 3 = 90, 8*x* + 2 = 90, 8*x* = 88, *x* = 11. The measure of angle 2 is 5(11) − 3 = 55 − 3 = 52°.

**20. b.** Angles 3 and 7 are vertical angles, so their measures are equal. Angles 4 and 8 are also vertical angles, so their measures are also equal. Since angles 4 and 7 are congruent, and angles 4 and 8 are congruent, angles 7 and 8 must be congruent. In the same way, since angles 4 and 7 are congruent and angles 3 and 7 are congruent, angles 3 and 4 must be congruent. Angles 3, 4, 7, and 8 are all congruent. In fact, since angles 3 and 4 are complementary and angles 7 and 8 are complementary (since lines *CD* and *GH* are perpendicular) all four angles measure 45°. However, nothing is known about angles 1, 2, 5, and 6. It cannot be stated that angle 2 = angle 3.

**21. e.** Since lines *EF* and *AB* are perpendicular, angle 8 measures 90°. Therefore, 90 + angle 9 = 133°, 133 − 90 = 43, so angle 9 measures 43°. Angles 3 and 9 form a line; the sum of their measures is 180°. The measure of angle 3 is equal to: 180 − 43 = 137°.

**22. d.** Angles 21, 17, and 11 are alternating angles; their measure are equal. Since angle 11 and angle 5 are supplementary (they combine to form a line), angles 21 and 5 must be supplementary. Therefore, $x^2 + 11 = 180$, $x^2 = 169$, $(x + 13)(x - 13) = 0$, and *x* = 13 (disregard the negative value of *x*, since every numbered angle is greater than 0). If *x* = 13, then the measure of angle 3 is 9(13) + 1 = 117 + 1 = 118°. Since angle 3 and angle 10 are vertical angles, angle 10 is also 118°.

**23. c.** Angles 1 and 8 are both right angles. In addition to be equal, they are also supplementary, since 90 + 90 = 180°. Angles 19, 15, and 20 also add to 180°, since these angles form a line. Angle 1 + angle 8 = angle 19 + angle 15 + angle 20 = 180°.

**24. d.** Angles 5, 12, 16, and 22 are alternating (vertical) angles; their measures are equal; 8*x* − 4 = 7*x* + 11, *x* − 4 = 11, *x* = 15; 8(15) − 4 = 120 − 4 = 116°. Notice that you could also substitute 15 into the measure of angle 22: 7(15) + 11 = 105 + 11 = 116°. The measure of angles 5, 12, 16, and 22 is 116°.

**25. b.** Since lines $AB$ and $GH$ are perpendicular, angle 20 measures 90° and the sum of the measures of angles 14 and 15 is 90°. Therefore, angle $14 = 15x + 6 - 90$ and angle $15 = 18x - 90$. Since the sum of angles 14 and 15 is 90°, $(15x + 6 - 90) + (18x - 90) = 90$, $33x - 174 = 90$, $33x = 264$, $x = 8$. Therefore, the sum of angles 20 and 14 is $15(8) + 6 = 126$, and the measure of angle 14 is equal to: $126 - 90 = 36$. Since angles 14 and 19 are vertical angles, angle 19 also measures 36°.

# ▶ Chapter 10

**1. b.** The measures of the angles of a triangle add to 180°. Therefore, $3x + 4x + 5x = 180$, $12x = 180$, and $x = 15$.

**2. e.** The measures of the angles of a triangle add to 180°. Therefore, $5x + 10 + x + 10 + 2x = 180$, $8x + 20 = 180$, $8x = 160$, and $x = 20$. The measure of angle $A$ is $5(20) + 10 = 110$, the measure of angle $B$ is $(20) + 10 = 30$, and the measure of angle $C$ is $2(20) = 40$. Since the largest angle of triangle $ABC$ is greater than 90° and no two angles of the triangle are equal in measure, triangle $ABC$ is obtuse and scalene.

**3. c.** An angle and its exterior angle are supplementary. Therefore, $8x + 16x + 12 = 180$, $24x + 12 = 180$, $24x = 168$, $x = 7$. Since $x = 7$, the measure of angle $F = 8(7) = 56°$.

**4. e.** The measures of the angles of a triangle add to 180°. Therefore, $2x + 5 + 2x + 5 + 3x - 5 = 180$, $7x + 5 = 180$, $7x = 175$, and $x = 25$. The measure of angle $A$ is $2(25) + 5 = 50 + 5 = 55$. Since an angle and its exterior angle are supplementary, the measure of an angle exterior to $A$ is $180 - 55 = 125°$.

**5. e.** Since an angle exterior to angle $F$ is 120°, the measure of interior angle $F$ is 60°, and the sum of the measures of interior angles $D$ and $E$ is 120°. Angles $D$ and $E$ could each measure 60°, making triangle $DEF$ acute and equilateral, but these angles could also measure 100° and 20° respectively, making triangle $DEF$ obtuse and scalene.

However, triangle $DEF$ cannot be isosceles. Angle $F$ measures 60°; if either angles $D$ or $E$ measure 60°, the other must also measure 60°, making triangle $DEF$ equilateral. Angles $D$ and $E$ cannot be congruent to each other without also being congruent to angle $F$. Therefore, triangle $DEF$ can be acute, obtuse, scalene, or equilateral, but not isosceles.

**6. a.** The measure of an exterior angle is equal to the sum of the measures of the interior angles to which the exterior angle is not adjacent. Therefore, the measure of angle 5 is equal to the sum of the measures of angles 1 and 3, not angles 1 and 2. It is true that the sum of angles 4 and 1 is equal to the sum of angles 3 and 6, since both pairs of angles form lines. It is also true that the sum of angles 2 and 3 is equal to the measure of angle 4, since angle 4 is an exterior angle that is not adjacent to 2 or 3. Since there are 180° in a triangle, the sum of angles 1, 2, and 3 is equal to 180°. The sum of the measures of one exterior angle from each vertex of a triangle is 360°, so the statement in choice **e** is also true.

**7. d.** Angles 3 and 6 are supplementary. Therefore, the measure of angle $3 = 180 - 115 = 65$. The measure of an exterior angle is equal to the sum of the measures of the interior angles to which the exterior angle is not adjacent. Therefore, the sum of angles 2 and 3 is equal to the measure of angle 4: $75 + 65 = 140°$.

**8. c.** If one exterior angle is taken from each vertex of a triangle, the sum of these exterior angles is 360°; $7x + 2 + 8x + 8x - 10 = 23x - 8 = 360$, $23x = 368$, $x = 16$. Therefore, angle 4 measures $7(16) + 2 = 112 + 2 = 114$. An exterior angle and its adjacent interior angle are supplementary, so the measure of angle 1 is equal to: $180 - 114 = 66°$.

**9. a.** Since there are 180° in a triangle, $x^2 + 1 + 9x - 7 + 6x + 2 = 180$, $x^2 + 15x - 4 = 180$, $x^2 + 15x - 184 = 0$, $(x - 8)(x + 23) = 0$, $x = 8$ (disregard the negative value of $x$ since an angle cannot have a negative measure). Therefore, the measure of angle 1 is $(8)2 + 1 = 64 + 1 = 65$, the measure of angle 2 is $9(8) - 7 = 72 - 7 = 65$, and the measure of angle 3 is $6(8) + 2 = 48 + 2 = 50°$. Since exactly two of the angles of triangle $ABC$ are equal, triangle $ABC$ is isosceles.

**10. d.** If the measure of angle $F$ is $x$, then the sum of the measures of angles $D$ and $E$ is $2x$. Since there are $180°$ in a triangle, $x + 2x = 180, 3x = 180$, and $x = 60$. Since the measure of an exterior angle is equal to the sum of the measures of the interior angles to which the exterior angle is not adjacent, the measure of an angle exterior to $F$ is equal to $2(60) = 120°$.

**11. d.** The measures of corresponding sides of similar triangles are in the same ratio. The ratio of side $AB$ to side $DE$ is equal to the ratio of side $AC$ to side $DF$. Therefore, $\frac{90}{60} = \frac{72}{x}, \frac{3}{2} = \frac{72}{x}, 3x = 144, x = 48$. The length of $\overline{DF}$ is 48.

**12. b.** The measures of corresponding sides of similar triangles are in the same ratio. The ratio of side $AB$ to side $DE$ is equal to the ratio of side $AC$ to side $DF$. Therefore, $\frac{10x-2}{2x+2} = \frac{6x}{x+2}, (10x-2)(x+2) = (6x)(2x+2), 10x^2 - 2x + 20x - 4 = 12x^2 + 12x, 2x^2 - 6x + 4 = 0, x^2 - 3x + 2 = 0, (x-2)(x-1) = 0, x = 1; x = 2$. If $x = 1$, then side $DF = (1) + 2 = 3$. However, every side of triangle $DEF$ is greater than 3. Therefore, $x$ must be equal to 2; $2 + 2 = 4$. The length of side $DF$ is 4.

**13. b.** The measures of corresponding sides of similar triangles are in the same ratio. The ratio of side $BC$ to side $EF$ is $1:\frac{1}{5}$, or 5:1. Therefore, each side of triangle $ABC$ is five times the length of its corresponding side in triangle $DEF$, and each side of triangle $DEF$ is $\frac{1}{5}$ the length of its corresponding side in triangle $ABC$. Corresponding angles of similar triangles are congruent. Therefore, angle $A$ is congruent to angle $D$, not five times its measure.

**14. c.** Since triangles $JKL$ and $MNO$ are congruent and equilateral, every side of triangle $JKL$ is congruent every other side of triangle $JKL$ and congruent to every side of triangle $MNO$. Therefore, $6x + 3 = x^2 - 4, x^2 - 6x - 7, (x-7)(x+1) =$

$0, x = 7$. Disregard the negative value of $x$, since a side of a triangle cannot be negative. Every length of triangles $JKL$ and $MNO$ is equal to $6(7) + 3 = 42 + 3 = 45$.

**15. a.** Since the triangles are similar, the measures of their corresponding sides are in the same ratio. Sides $HG$ and $PQ$ are corresponding sides that are congruent. Therefore, the ratio of side $HG$ to side $PQ$ is 1:1. Each side of triangle $GHI$ is congruent to its corresponding side; therefore, side $GI$ is congruent to its corresponding side.

**16. d.** Angles 10 and 13 are vertical angles. Since vertical angles are equal, angle 10 is also $94°$. There are $180°$ in a triangle, and angles 6 and 7 are congruent. Therefore, $94 + x + x = 180, 2x = 86, x = 43$. Since the measure of angle 7 is 43 and angles 7 and 8 are supplementary, the measure of angle 8 is equal to $180 - 43 = 137°$.

**17. b.** The measure of an exterior angle is equal to the sum of the interior angles to which the exterior angle is not adjacent. Therefore, angle 3 is equal to the sum of angles 6 and 10: $10x + 15 = 8x - 3 + 3x + 7, 10x + 15 = 11x + 4, x = 11$. The measure of angle 3 is $10(11) + 15 = 110 + 15 = 125$. Since angles 3 and 7 are supplementary, the measure of angle 7 is $180 - 125 = 55°$.

**18. d.** Angles 10 and 13 are vertical angles; their measures are equal. Angle 5 and the sum of angles 10 and 11 are equal, since angle 5 and the combination of angles 10 and 11 are alternating angles. Therefore, $10x + 2 = 4x - 4 + 7x - 6, 10x + 2 = 11x - 10, x = 12$. Since the measure of an exterior angle is equal to the sum of the interior angles to which the exterior angle is not adjacent, angle 5 is equal to the sum of angles 10 and 7: $10(12) + 2 = 120 + 2 = 122°$.

**19. c.** If the measure of angle 14 is $x$, then the measure of angle 11 is $2x$ and the measure of angle 8 is $2.5x$. Since angles 8 and 11 are supplementary, $2x + 2.5x = 180, 4.5x = 180, x = 40$. Therefore, angle 14 $= 40°$ and angle 11 $= 2(40) = 80°$. Since angles 10, 11, and 14 form a line, the measure of angle 10 is equal to $180 - (40 + 80) = 180 - 120 = 60°$.

**20. a.** If angle 8 is greater than angle 7, angle 7 cannot be the right angle of the triangle; otherwise, angles 8 and 7 would be congruent. Therefore, angle 7 is one of the acute, congruent angles of the right triangle, which means that it measures 45°. Since angles 8 and 5 are congruent, angle 6 must be the other 45° angle of the triangle, making angle 10 the right angle of the triangle. Since angle 6 is 45° and angles 2 and 6 are supplementary, the measure of angle 2 is $180 - 45 = 135°$. Since angles 6 and 9 are alternating angles, the measure of angle 9 is 45°, and since angles 7 and 12 are alternating angles, the measure of angle 12 is also 45°. Therefore, the sum of angles 9 and 12 is $45 + 45 = 90$, not 135, so angle 2 is not equal to angle 9 + angle 12.

## ▶ Chapter 11

**1. e.** Use the Pythagorean theorem: $(x-3)^2 + (x+4)^2 = (2x-3)^2$, $x^2 - 6x + 9 + x^2 + 8x + 16 = 4x^2 - 12x + 9$, $2x^2 + 2x + 25 = 4x^2 - 12x + 9$, $2x^2 - 14x - 16 = 0$, $x^2 - 7x - 8 = 0$, $(x-8)(x+1) = 0$, $x = 8$. Disregard the negative value of $x$, since a side of a triangle cannot be negative. Since $x = 8$, the length of the hypotenuse is $2(8) - 3 = 16 - 3 = 13$.

**2. d.** Use the Pythagorean theorem: $8^2 + x^2 = (8\sqrt{5})^2$, $64 + x^2 = 320$, $x^2 = 256$, $x = 16$ ($x$ cannot equal $-16$ since the side of a triangle cannot be negative).

**3. b.** Use the Pythagorean theorem: $9^2 + 15^2 = x^2$, $81 + 225 = x^2$, $x^2 = 306$, $x = \sqrt{306} = 3\sqrt{34}$.

**4. c.** If the shorter base is $a$, then the longer base is $3a$. Use the Pythagorean theorem: $a^2 + (3a)^2 = c^2$, $a^2 + 9a^2 = c^2$, $10a^2 = c^2$, $c = a\sqrt{10}$.

**5. b.** The hypotenuse of the given triangle is $x + 5$, since $x$ cannot be 0 or negative (the side of a triangle cannot be 0 or negative); therefore, $x + 5 > x > x - 5$. Using the Pythagorean theorem, $(x)^2 + (x-5)^2 = (x+5)^2$; $x^2 + x^2 - 10x + 25 = x^2 + 10x + 25$, $2x^2 - 10x + 25 = x^2 + 10x + 25$, $x^2 - 20x = 0$, $x(x - 20) = 0$, and $x = 20$, since $x$

cannot be 0. Therefore, the sides of the triangle measure 20, $20 - 5 = 15$, and $20 + 5 = 25$. The sides of the triangle, 15, 20, 25, are a multiple of the triangle with sides 3, 4, 5, since $3(5) = 15$, $4(5) = 20$, and $5(5) = 25$.

**6. b.** If the measure of one base angle of a right triangle is 45°, then the measure of the other base angle must also be 45° ($180 - (90 + 45) = 45$). Therefore, $TUV$ is an isosceles right triangle, or a 45-45-90 right triangle. The hypotenuse of a 45-45-90 right triangle is equal to $\sqrt{2}$ times the length of a base. Since the length of a base is 10, the hypotenuse of the triangle is $10\sqrt{2}$ units.

**7. a.** The hypotenuse of an isosceles right triangle is equal to $\sqrt{2}$ times the length of a base, and the base of the triangle is equal to $\frac{\sqrt{2}}{2}$ times the length of the hypotenuse. Since the hypotenuse of the triangle is $x$, the bases of the triangle each measure $\frac{x\sqrt{2}}{2}$ units.

**8. d.** If angle $Q$ is the right angle of the triangle, then side $QR$ is a base of the right triangle. If the sine of $P$ is equal to the sine of $R$, then angle $P$ is equal to angle $R$. Therefore, $PQR$ is an isosceles right triangle. The hypotenuse of an isosceles right triangle is equal to $\sqrt{2}$ times the length of a base. Since the length of a base is 4, the hypotenuse of the triangle is $4\sqrt{2}$ units.

**9. e.** To find the hypotenuse of a right triangle, you must add the squares of the base lengths. However, only the length of one base is given. There is no information provided about the angles of triangle $IJK$, so you cannot determine if this is a certain type of right triangle. Therefore, the length of side $IK$ cannot be determined from the information provided.

**10. d.** Since angle $IKD$ is 135°, and it is supplementary to angle $IKJ$, the measure of angle $IKJ = 180 - 135 = 45$. Lines $GH$ and $CD$ form a right angle, $IJK$, since these lines are perpendicular. Since angle $IKJ = 45°$, angle $JIK$ is also 45° ($180 - (90 + 45) = 45$). Therefore, triangle $IJK$ is an isosceles right triangle. The length of side $IJ$ is

25 units, since line *CD* is 25 units from line *AB*. The hypotenuse of an isosceles right triangle is equal to $\sqrt{2}$ times the length of a base. Since the length of a base, *IJ*, is 25, the hypotenuse, *IK*, is $25\sqrt{2}$ units.

**11. e.** If angle *A* is twice the measure of angle *C*, then let angle *C* = *x* and let angle *A* = 2*x*; *x* + 2*x* + 90 = 180, 3*x* = 90, *x* = 30. The measure of angle *C* is 30° and the measure of angle *A* is 2(30) = 60°. Therefore, this is a 30-60-90 right triangle. Side *AB* is the shorter base, since it is the side opposite the smallest angle. The hypotenuse of a 30-60-90 right triangle is twice the length of the shorter base. Therefore, the measure of *AC* is 2(7) = 14 units.

**12. d.** Since angle *B* of triangle *ABC* is a right angle and angle *C* is 60°, angle *A* must be 30° (180 − (90 + 60) = 30). Therefore, this is a 30-60-90 right triangle. Side *AB* is the longer base, since it is the side opposite the larger base angle. The measure of the larger base of a 30-60-90 right triangle is $\sqrt{3}$ times the length of the shorter base. Therefore, the length of side *BC* is equal to the length of side *AB* divided by $\sqrt{3}$: $\frac{9}{\sqrt{3}} = \frac{9\sqrt{3}}{3} = 3\sqrt{3}$ units.

**13. c.** Since angle *B* of triangle *ABC* is a right angle and angle *C* is 60°, angle *A* must be 30° (180 − (90 + 60) = 30). Therefore, this is a 30-60-90 right triangle. Side *AB* is the longer base, since it is the side opposite the larger base angle. The hypotenuse of a 30-60-90 right triangle is twice the length of the shorter base. Therefore, the measure of side *BC* is $\frac{(6x+2)}{2} = 3x + 1$ units. The measure of the larger base of a 30-60-90 right triangle is $\sqrt{3}$ times the length of the shorter base. Therefore, the length of side *AB* is $(3x + 1)\sqrt{3}$ units.

**14. c.** Since angle *B* of triangle *ABC* is a right angle and angle *C* is 60°, angle *A* must be 30° (180 − (90 + 60) = 30). Therefore, this is a 30-60-90 right triangle. Side *AB* is the longer base, since it is the side opposite the larger base angle. The hypotenuse of a 30-60-90 right triangle is twice the length of the shorter base. Therefore, if the length of side *BC* is *x*, then the length of side *AC* is 2*x*, and their sum is equal to 3*x*; 3*x* = 12, *x* = 4. The length of side *BC* is 4 units, and the length of side *AB* is $4\sqrt{3}$ units, since the longer base of a 30-60-90 right triangle is $\sqrt{3}$ times the length of the shorter base.

**15. b.** Side *AB* of equilateral triangle *ABC* is 8 units; therefore, side *AC* is also 8 units. Since side *AE* is 20 units, side *CE* must be 20 − 8 = 12 units. Angle *ACB* is 60°, since every angle of an equilateral triangle is 60°. Angles *ACB* and *ECD* are vertical angles, so angle *ECD* also measures 60°. Since angle *ECD* is 60°, angle *CED* is 30°, and triangle *CDE* is a 30-60-90 right triangle. Side *CD* is half the length of side *CE*, since it is the shorter base of the 30-60-90 right triangle. Therefore, side *CD* is equal to $\frac{12}{2} = 6$. Line segment *BC* is 8 units, since it is a side of equilateral triangle *ABC*, of which every side measures 8 units. The length of side *BD* is equal to the sum of side *BC* and side *CD*: 8 + 6 = 14 units.

**16. c.** Since triangle *ABC* is an isosceles right triangle, the measures of both bases are equal. Therefore, the tangent of either angle *A* or angle *C* is the tangent of 45°, which is 1, since any number divided by itself is 1.

**17. a.** The tangent of an angle is equal to the side opposite the angle divided by the side adjacent to the angle; $\frac{\overline{BC}}{\overline{AB}} = \frac{(x^2 - 2x)}{(3x + 6)} = \frac{x(x - 2)}{3(x - 2)} = \frac{x}{3}$.

**18. a.** The sine of one base angle of a right triangle is equal to the cosine of the other base angle. Therefore, the sine of angle *C* is equal to the cosine of angle *A*, and the cosine of angle *C* is equal to the sine of angle *A*. Why? If the cosine of angle *C* is $\frac{x}{y}$, then the length of side *BC* is *x* and the length of side *AC* is *y*. Since the sine of angle *A* is equal to the side opposite angle *A*, side *BC*, divided by the hypotenuse, side *AC*, the sine of angle *A* is also $\frac{x}{y}$.

**19. a.** If the sine of angle *A* is $\frac{15}{17}$, then the length of side *BC* is 15 and the length of side *AC* is 17, since the sine of an angle is equal to the side opposite the angle divided by the hypotenuse. Use the Pythagorean theorem to find the length of $\overline{AB}$: $(\overline{AB})^2 + 15^2 = 17^2$, $(\overline{AB})^2 + 225 = 289$, $(\overline{AB})^2 = 64$, $\overline{AB} = 8$. Since the cosine of an angle is equal to the side adjacent to the angle divided by the hypotenuse, the cosine of angle $A$ is equal to $\frac{\overline{AB}}{\overline{AC}} = \frac{8}{17}$.

**20. e.** The tangent of angle $A$ is equal to 0.75, or $\frac{3}{4}$. Since the tangent of $A$ is the measure of side $BC$ divided by side $AB$, the ratio of side $BC$ to side $AB$ is 3:4. To find the relative length of side $AC$, use the Pythagorean theorem: $3^2 + 4^2 = (\overline{AC})^2$, $9 + 16 = (\overline{AC})^2$, $25 = (\overline{AC})^2$, $\overline{AC} = 5$. The ratio of side $BC$ to side $AB$ to side $AC$ is 3:4:5. Since the length of side $AB$ is 4 less, not 1 less, than the length of side $AC$, multiply each number in the ratio by 4: $\overline{BC} = 3(4) = 12$ units, $\overline{AB} = 4(4) = 16$ units, and $\overline{AC} = 5(4) = 20$ units. If the length of $\overline{AB}$ is 16 units and the length of $\overline{AC}$ is 20 units, $\overline{AB}$ is 4 less than $\overline{AC}$, and the tangent of angle $A$ is still 0.75, since $\frac{12}{16} = 0.75$. The length of $\overline{BC}$ is 12 units.

**21. e.** Since $\overline{KM} = 10$ and $\overline{LM} = 5$, the cosine of angle $LMK$ is $\frac{\overline{LM}}{\overline{KM}} = \frac{1}{2}$. The cosine of a 60° angle is $\frac{1}{2}$, so the measure of angle $LMK$ is 60°. Angles $LMK$ and $KIJ$ are alternating angles, so their measures are equal. Therefore, angle $KIJ$ is also 60°.

**22. d.** Angles $IKJ$ and $LKM$ are vertical angles; their measures are equal. Therefore, each angle is equal to $\frac{60}{2} = 30°$. The sine of a 30° angle is $\frac{1}{2}$. Since the measure of angle $LKM$ is 30°, the length of $\overline{LM}$ divided by the length of $\overline{KM}$ is equal to $\frac{1}{2}$: $\frac{\overline{LM}}{\overline{KM}} = \frac{1}{2}$, $\frac{8}{\overline{KM}} = \frac{1}{2}$, $\overline{KM} = 16$ units.

**23. a.** The tangent of a 60° angle is $\sqrt{3}$; therefore, the measure of angle $JIK$ is 60°. Angles $JIK$ and $LMK$ are alternating angles; their measures are equal. Angle $LMK$ is 60° and its tangent is $\sqrt{3}$. Therefore, the measure of $\overline{LK}$, the side opposite angle $LMK$, is $\sqrt{3}$ times the length of $\overline{LM}$, the side adjacent to angle $LMK$.

**24. b.** Angles $KMH$ and $KML$ are supplementary angles; their measures add to 180°. If the measure of angle $KML$ is $y$ and the measure of angle

$KMH$ is $3y$, then $4y = 180$, $y = 45°$. The measure of angle $KML$ is 45°. Angle $JIK$ is also 45°, since $KML$ and $JIK$ are alternating angles. Since $JIK$ is 45° and $KJI$ is 90°, angle $IKJ$ is 45° ($180 - (45 + 90) = 45$). Therefore, triangle $IJK$ is a 45-45-90 right triangle. The hypotenuse of a 45-45-90 right triangle is $\sqrt{2}$ times the length of either base: $(x\sqrt{6})(\sqrt{2}) = x\sqrt{12} = x\sqrt{4}\sqrt{3} = 2x\sqrt{3}$ units.

**25. d.** Angles $IKJ$ and $LKM$ are vertical angles; their measures are equal. Since the sine of an angle is equal to the length of the side opposite the angle divided by the length of the hypotenuse, the sine of angle $IKJ$ is equal to $\frac{\overline{IJ}}{\overline{IK}}$ and the sine of angle $LKM$ is equal to $\frac{\overline{LM}}{\overline{KM}}$. Two equal angles have equal sines; $\frac{\overline{IJ}}{\overline{IK}} = \frac{\overline{LM}}{\overline{KM}}$, $\frac{2x-2}{2x+1} = \frac{2x+2}{3x-1}$, $6x^2 - 6x - 2x + 2 = 4x^2 + 2x + 4x + 2$, $6x^2 - 8x + 2 = 4x^2 + 6x + 2$, $2x^2 - 14x = 0$, $x^2 - 7x = 0$, $x(x - 7) = 0$, $x = 7$ ($x$ cannot be 0 since the length of a line cannot be negative—if $x$ were 0, $\overline{IJ}$ would be –2 units). If $x = 7$, then the length of $\overline{LM} = 2(7) + 2 = 14 + 2 = 16$ units.

## ▶ Chapter 12

**1. b.** Regular pentagons are equilateral; every side is equal in length. Therefore, every angle is equal in size, and every regular pentagon is similar to every other regular pentagon. However, regular pentagons are not necessarily congruent. One regular pentagon could be ten times the size of another. Heather's regular pentagons may not be congruent.

**2. b.** The sum of the exterior angles of any polygon is 360°. Therefore, the sum of the interior angles of this polygon is $(360)(3) = 1,080$. The sum of the interior angles of a polygon is equal to $180(s - 2)$, where $s$ is the number of sides of the polygon; $1,080 = 180(s - 2)$, $1,080 = 180s - 360$, $1,440 = 180s$, $144 = 18s$, $s = 8$. The polygon has 8 sides.

**3. a.** The sum of the interior angles of a polygon is equal to $180(s - 2)$, where $s$ is the number of sides of the polygon. If $x$ is the number of sides of the polygon, and the sum of the interior angles is 60 times that number, then $60x = 180(x - 2)$, $60x = 180x - 360$, $120x = 360$, $x = 3$. Andrea's polygon has three sides.

**4. c.** The sum of the exterior angles of any polygon is 360°. The sum of the interior angles of a polygon is equal to $180(s - 2)$, where $s$ is the number of sides of the polygon. If these sums are equal, then $180(s - 2) = 360$, $180s - 360 = 360$, $180s = 720$, and $s = 4$. The polygon has 4 sides.

**5. b.** The sum of the interior angles of a polygon is equal to $180(s - 2)$, where $s$ is the number of sides of the polygon. Since the number of sides is three less than $x$ and the sum of the interior angles is $9x^2$, $180(x - 3 - 2) = 9x^2$, $180x - 900 = 9x^2$, $9x^2 - 180x + 900 = 0$, $x^2 - 20x + 100 = 0$, $(x - 10)(x - 10) = 0$, $x = 10$. Therefore, the number of sides of the polygon is $(10) - 3 = 7$ sides.

**6. a.** The ratio of $\overline{AB}$ to $\overline{FG}$ is 4:1. Therefore, $\frac{4}{1} = \frac{4x + 4}{\overline{FG}}$, $4(\overline{FG}) = 4x + 4$, $\overline{FG} = x + 1$.

**7. e.** Although it is known that $ABCD$ and $EFGH$ are similar, the ratio of their corresponding sides is not provided, nor is the ratio of their perimeters. Therefore, the perimeter of $EFGH$ cannot be determined.

**8. d.** The number of sides of a polygon determines the sum of the interior angles of the polygon. The sum of the interior angles of any octagon is $180(8 - 2) = 180(6) = 1{,}080°$. These two octagons are not necessarily similar or congruent, although they could be. The ratio of $\overline{AB}$ to $\overline{ST}$ could be 1:1, but there could also be no ratio between the sides of $ABCDEFGH$ and $STUVWXYZ$. No one side of $ABCDEFGH$ must equal one side of $STUVWXYZ$.

**9. c.** Since $\overline{AB}$ and $\overline{GH}$ are corresponding sides of similar polygons, the ratio of $\overline{GH}$ to $\overline{AB}$ is equal to the ratio of the perimeter of $GHIJKL$ to the perimeter of $ABCDEF$; $GH{:}AB = \frac{8x + 4x}{12x + 6x} = \frac{12x}{18x} = \frac{2}{3} = 2{:}3$.

**10. e.** Since $\overline{AB}$ and $\overline{TU}$ are corresponding sides of similar polygons, the ratio of $\overline{AB}$ to $\overline{TU}$ is equal to the ratio of the perimeter of $ABCDE$ to the perimeter of $TUVWX$; $\overline{AB}{:}\overline{TU} = \frac{5x - 1}{4x - 2} = \frac{4}{3}$, $15x - 3 = 16x - 8$, $-3 = x - 8$, $x = 5$. Therefore, the length of $\overline{AB}$ is $5(5) - 1 = 25 - 1 = 24$ units. Since $ABCDE$ is a regular polygon with 5 sides, each of the 5 sides measures 24 units. Therefore, the perimeter of $ABCDE$ is $(5)(24) = 120$ units.

**11. d.** The perimeter of $ABCD$ is equal to the sum of its sides: $2x + 2x + 2 + 3x + 3x - 2 = 60$, $10x = 60$, $x = 6$. Therefore, the length of $\overline{BC}$ is $3(6) - 2 = 18 - 2 = 16$ units.

**12. b.** If $x = 8$, then the perimeter of the figure is $11(8) - 4 = 88 - 4 = 84$. Every side of a regular polygon is equal in length. Therefore, the length of one side of this seven-sided polygon is $\frac{84}{7} = 12$ units.

**13. e.** If the perimeter of a regular pentagon is 75 units, then the length of one side of the pentagon is $\frac{75}{5} = 15$ units. The ratio of a length of $ABCDE$ to a length of $PQRSTU$ is 5:6. Therefore, if $x$ is the length of a side of $PQRSTU$, then $\frac{5}{6} = \frac{15}{x}$, $5x = 90$, and $x = 18$ units. Since the length of one side of the regular hexagon is 18, the perimeter of the hexagon is $(6)(18) = 108$ units.

**14. c.** The hypotenuse of an isosceles right triangle is equal to $\sqrt{2}$ times the length of a base of the triangle. If the hypotenuse is $5\sqrt{2}$, then the length of each base is $\frac{(5\sqrt{2})}{(\sqrt{2})} = 5$. Therefore, the perimeter of the triangle is $5 + 5 + 5\sqrt{2} = 10 + 5\sqrt{2}$ units.

**15. a.** The sum of the interior angles of a polygon is equal to $180(s - 2)$, where $s$ is the number of sides of the polygon; $720 = 180(s - 2)$, $720 = 180s - 360$, $1{,}080 = 180s$, $108 = 18s$, $s = 6$. Since the polygon is regular, has 6 sides, and the length of one side is $3x^2$, the perimeter of the polygon is $6(3x^2) = 18x^2$ units.

# ▶ Chapter 13

**1. a.** Andrew's polygon has four sides, so it must be a quadrilateral. It does not contain a right angle, so it cannot be a rectangle or a square, since these quadrilaterals each have four right angles. However, this polygon could be a parallelogram or a rhombus, since these are four-sided polygons that do not necessarily contain a right angle.

**2. c.** Angles $E$ and $H$ are consecutive angles in rhombus $EFGH$. Therefore, their measures are supplementary: $3x + 5 + 4x = 180$, $7x + 5 = 180$, $7x = 175$, $x = 25$. The measure of angle $H$ is $4(25) = 100$. Since opposite angles of a rhombus are congruent and angles $H$ and $F$ are opposite angles, angle $F$ is also $100°$.

**3. e.** In order to form a square or a rhombus, you must have four identical line segments. Since nothing about the angles is stated in the question, these four line segments could be connected at right angles to form a square. Since a square is a type of rectangle and a type of rhombus, both of which are types of parallelograms, any of these four types of quadrilaterals could be formed.

**4. d.** Angles $DCE$ and $ECB$ are complementary angles; these angles combine to form right angle $DCB$ of the rectangle. These angles are only equal if $ABCD$ is a square. However, $ABCD$ is a rectangle, and not all rectangles are squares. Therefore, it is not always true that angle $DCE =$ angle $ECB$.

**5. b.** The diagonals of a rhombus are perpendicular. A rhombus has four congruent sides, but its diagonals are not necessarily congruent, and only the opposite angles of a rhombus are necessarily congruent. If a quadrilateral has perpendicular diagonals, it must have four congruent sides.

**6. e.** If $x$ is the width of the rectangle, then $2x - 4$ is the length of the rectangle. Since opposite sides of a rectangle are congruent, the perimeter of the rectangle is equal to $2x - 4 + x + 2x - 4 + x = 6x - 8$.

**7. a.** Every side of a rhombus is equal in length. One side of this rhombus is equal to $\frac{168}{4} = 42$ units. Therefore, $x^2 - 6 = 42$, $x^2 = 48$, $x = \sqrt{48} = \sqrt{16}\sqrt{3} = 4\sqrt{3}$.

**8. a.** If the square and the rectangle share a side, then the width of the rectangle is equal to the length of the square. If the length of the square is $x$, then $x + x + x + x = 2$, $4x = 2$, $x = \frac{1}{2}$. Since the width of the rectangle is equal to the length of the square and the length of the rectangle is 4 times the length of the square, the width of the rectangle is $\frac{1}{2}$ and the length is $4(\frac{1}{2}) = 2$. Therefore, the perimeter of the rectangle is equal to: $\frac{1}{2} + 2 + \frac{1}{2} + 2 = 5$ units.

**9. b.** Since the rhombus and the square have the same perimeter and both figures have four congruent sides, every side of the rhombus and every side of the square is equal to exactly one-fourth of the perimeter. Therefore, every side of the rhombus must be congruent to every side of the square.

**10. a.** The perimeter of each small square is $8x$ units; therefore, the length of a side of each small square is $\frac{8x}{4} = 2x$ units. Since the new, large square is comprised of two sides from each of the four squares (the remaining four sides are now within the large square), the perimeter of the new, large square is equal to: $4(2x + 2x) = 4(4x) = 16x$ units.

**11. d.** The diagonal of a square is the hypotenuse of an isosceles right triangle, since the 4 angles of a square are right angles, and the sides of a square are all congruent to each other. Therefore, the measure of a side of the square is equal to the measure of the diagonal divided by $\sqrt{2}$: $\frac{2x\sqrt{2}}{\sqrt{2}} = 2x$ units. Therefore, the perimeter of the square is equal to $2x + 2x + 2x + 2x = 8x$ units.

**12. e.** The tangent of angle $ACB$ is equal to the length of $\overline{AB}$ divided by the length of $\overline{BC}$. Therefore, if the length of $\overline{BC}$ is 8 units, then $\frac{\overline{AB}}{8} = 8$, and $\overline{AB} = 64$. The perimeter of $ABCD$ is equal to $64 + 8 + 64 + 8 = 144$ units.

**13. b.** The diagonal of a square is the hypotenuse of an isosceles right triangle, since the 4 angles of a square are right angles, and the sides of a square are all congruent to each other. Therefore, the measure of a side of the square is equal to the measure of the diagonal divided by $\sqrt{2}$: $\frac{(2x-2)\sqrt{2}}{\sqrt{2}} = 2x - 2$ units. Therefore, the perimeter of the square, $5x + 1$, is equal to $2x - 2 + 2x - 2 + 2x - 2 + 2x - 2 = 8x - 8$; $8x - 8 = 5x + 1$, $3x = 9$, $x = 3$. The length of a side of the square is equal to $2(3) - 2 = 6 - 2 = 4$ units.

**14. d.** The cosine of $ACD$ is equal to the length of $\overline{CD}$ divided by the length of diagonal $AC$. Diagonal $AC$ along with sides $AD$ and $DC$ form a right triangle. Use the Pythagorean theorem to find the length of $\overline{AD}$: $(\overline{AD})^2 + 12^2 = 13^2$, $(\overline{AD})^2 + 144 = 169$, $(\overline{AD})^2 = 25$, $\overline{AD} = 5$ units. If the length of $\overline{AD}$ is 5 units, then the length of $\overline{BC}$ is also 5 units. Since the length of $\overline{CD}$ is 12 units, the length of $\overline{AB}$ is also 12 units. The perimeter of $ABCD$ is $5 + 12 + 5 + 12 = 34$ units. Notice that this is only one possible perimeter of $ABCD$. The cosine could have been given in reduced form. The length of $\overline{CD}$ could be 24, and the length of $\overline{AC}$ could be 26. However, since the length of every side of $ABCD$ is an integer, the only possible perimeters are multiples of 34. Choice **d** is the only answer choice that is a multiple of 34.

**15. e.** The diagonals of a rhombus divide the rhombus into 4 congruent right triangles. Since angle $A$ measures 120°, angle $D$ also measures 120° and angles $B$ and $C$ each measure 60°. Each right triangle is made up of a right angle, a 30° angle (the angle formed by the diagonal bisecting the 60° angle of the rhombus), and a 60° angle (the angle formed by the diagonal bisecting the 120° angle of the rhombus). The bases of each triangle are made up of half the shorter diagonal (opposite the 30° angle) and half the longer diagonal (opposite the 60° angle). The hypotenuse is the side of the rhombus, 10 units. Since these right triangles are 30-60-90 right

triangles, the measure of the shorter base is $\frac{10}{2} = 5$ units, and the measure of the longer base is $5\sqrt{3}$ units. The longer base is half the length of the longer diagonal; therefore, the length of the longer diagonal is $2(5\sqrt{3}) = 10\sqrt{3}$ units.

## ▶ Chapter 14

**1. b.** If the base the triangle is $b$, then the height of the triangle is $\frac{1}{2}b$. The area of a triangle is equal to $\frac{1}{2}bh$. Therefore, the area of this triangle is equal to $\frac{1}{2}(b(\frac{1}{2}b)) = \frac{1}{4}b^2$.

**2. b.** The hypotenuse of an isosceles right triangle is $\sqrt{2}$ times the length of a base of the triangle. Therefore, a base of this triangle measures $\frac{x\sqrt{6}}{\sqrt{2}} = x\sqrt{3}$ units. The length of one base of the isosceles right triangle is its height, so the area of the triangle is $\frac{1}{2}(x\sqrt{3})(x\sqrt{3}) = \frac{3}{2}x^2$ square units.

**3. b.** $ABCD$ is a rectangle, which means that angle $B$ is a right angle, and triangle $EBC$ is a right triangle. Since $\overline{AB} = 30$ and $\overline{AE} = \overline{EB}$, then $\overline{EB} = \frac{30}{2} = 15$ units. The length of $\overline{BC}$ can be found using the Pythagorean theorem: $15^2 + (\overline{BC})^2 = 17^2$, $225 + (\overline{BC})^2 = 289$, $(\overline{BC})^2 = 64$, $\overline{BC} = 8$ units. Since $AB = 30$ units, $\overline{DC}$ also equals 30 units, since alternate sides of rectangles are congruent. The base of triangle $DEC$ is 30 units and the height of triangle $DEC$ is 8 units. Therefore, the area of triangle $DEC = \frac{1}{2}(30)(8) = \frac{1}{2}(240) = 120$ square units.

**4. a.** Every side of an equilateral triangle is equal in length, so the length of one side of this triangle is $\frac{36}{3} = 12$ units. If a line is drawn from the vertex of an angle of the triangle to its opposite base, this line represents the height of the triangle. This line cuts the triangle into two identical 30-60-90 right triangles, since this line is perpendicular to the base, and every angle of an equilateral triangle is 60°. The length of this line is equal to the length of the longer base of a 30-60-90 right triangle, which is $\sqrt{3}$ times the length of the shorter base. The shorter base is equal to half the length of the hypotenuse. Since the hypotenuse is 12 units, the length of the shorter base is 6 units and the height is $6\sqrt{3}$ units. Therefore, the area of the equilateral triangle is $\frac{1}{2}(12)(6\sqrt{3}) = (6)(6\sqrt{3}) = 36\sqrt{3}$ square units.

**5. e.** The area of the square is $x^2$ square units, since each side of the square has a length of $x$ units. Since the height of the triangle is $\frac{2}{3}x$ and the base of the triangle is $x$, the area of the triangle is equal to: $\frac{1}{2}(x)(\frac{2}{3})x = \frac{1}{3}x^2$ square units. Therefore, the size of the shaded area is equal to $x^2 - \frac{1}{3}x^2 = \frac{2}{3}x^2$ square units.

**6. c.** The area of a rectangle is equal to its length times its width. Therefore, the width of the rectangle is equal to its area divided by its length: $\frac{(x^2 + 7x + 10)}{(x + 2)}$; $x^2 + 7x + 10$ can be factored into $(x + 2)(x + 5)$. Cancel the $(x + 2)$ terms from the numerator and denominator of the fraction. The width of the rectangle is $x + 5$ units.

**7. a.** The area of a square is equal to the length of one side of the square multiplied by itself. Therefore, if the length of a side of a square is $x$, the area of the square is $x^2$. If the sides of the square are halved, then the area of the square becomes $(\frac{1}{2}x)(\frac{1}{2}x) = \frac{1}{4}x^2$. The area of the new square, $\frac{1}{4}x^2$, is one-fourth the area of the old square, $x^2$.

**8. d.** The area of a square is equal to the length of one side of the square multiplied by itself. Therefore, the length of one side of this square is equal to the square root of the area: $\sqrt{25} = 5$ units. Since the length of every side of a square is the same, the perimeter of this square is $(4)(5) = 20$ units. Set $3x - 4$ equal to 20 and solve for $x$: $3x - 4 = 20$, $3x = 24$, $x = 8$.

**9. c.** If the width of the rectangle is $x$, then the length of the rectangle is $3x - 2$. Since the area of a rectangle is equal to its length times its width, $(x)(3x - 2) = 96$, $3x^2 - 2x = 96$, $3x^2 - 2x - 96 = 0$, $(3x + 16)(x - 6) = 0$, and $x = 6$ (disregard the negative value of $x$ since the width of rectangle cannot be negative. Since $x = 6$, the length of the rectangle is equal to $3(6) - 2 = 18 - 2 = 16$ cm.

**10. d.** Angle $ACD$ is the 30° angle of 30-60-90 right triangle $ACD$. The sine of 30° is $\frac{1}{2}$. Since the sine of an angle is equal to the side opposite the angle divided by the hypotenuse, $\frac{1}{2} = \frac{\overline{AD}}{20}$, $2(\overline{AD}) = 20$, $\overline{AD} = 10$ units. $\overline{AD}$ is the side of the right triangle opposite the 30° angle; therefore, it is the shorter side of the right triangle. $\overline{DC}$, the longer side of the triangle, is $\sqrt{3}$ times the length of $\overline{AD}$: $10\sqrt{3}$. Since the area of a rectangle is equal to its length times its width, the area of $ABCD = (10)(10\sqrt{3}) = 100\sqrt{3}$ square units.

**11. a.** The volume of a cylinder is equal to $\pi r^2 h$, where $r$ is the radius of the cylinder and $h$ is the height of the cylinder. Only the values of the radius and height given in choice a hold true in the formula: $\pi(3)^2(5) = \pi(9)(5) = 45\pi$ in.$^3$.

**12. d.** The volume of a cylinder is equal to $\pi r^2 h$, where $r$ is the radius of the cylinder and $h$ is the height of the cylinder. Since Terri's glass is only $\frac{2}{3}$ full, the height of the water is $\frac{2}{3}(15) = 10$ cm. Therefore, the volume of water is equal to: $\pi(2)^2(10) = 40\pi$ cm$^3$.

**13. b.** The volume of a cylinder is equal to $\pi r^2 h$, where $r$ is the radius of the cylinder and $h$ is the height of the cylinder. If the volume of cylinder A is $\pi r^2 h$, then the volume of cylinder B is $\pi(\frac{1}{3}r)^2(3h) = \pi(\frac{1}{9}r^2)(3h) = \pi\frac{1}{3}r^2 h$, which is $\frac{1}{3}$ the volume of cylinder A.

**14. d.** The volume of a cylinder is equal to $\pi r^2 h$, where $r$ is the radius of the cylinder and $h$ is the height of the cylinder. The volume of this cylinder is equal to $\pi(2x)^2(8x + 2) = \pi 4x^2(8x + 2) = (32x^3 + 8x^2)\pi$.

**15. a.** The volume of a cylinder is equal to $\pi r^2 h$, where $r$ is the radius of the cylinder and $h$ is the height of the cylinder. Since the height is 4 times the radius, the volume of this cylinder is equal to $\pi r^2(4r) = 256\pi$, $4r^3 = 256$, $r^3 = 64$, $r = 4$. The radius of the cylinder is 4 cm.

**16. e.** The volume of a rectangular solid is equal to $lwh$, where $l$ is the length of the solid, $w$ is the width of the solid, and $h$ is the height of the solid. If $x$ represents the width (and therefore, the height as well), then the length of the solid is equal to $2(x + x)$, or $2(2x) = 4x$. Therefore, $(4x)(x)(x) = 108$, $4x^3 = 108$, $x^3 = 27$, and $x = 3$. If the width and height of the solid are each 3 in, then the length of the solid is $2(3 + 3) = 2(6) = 12$ in.

**17. c.** One face of a cube is a square. The area of a square is equal to the length of one side of the square multiplied by itself. Therefore, the length of a side of this square (and edge of the cube) is equal to $\sqrt{9x}$, or $3\sqrt{x}$ units. Since every edge of a cube is equal in length and the volume of a cube is equal to $e^3$, where $e$ is the length of an edge (or $lwh$, $l$ is the length of the cube, $w$ is the width of the cube, and $h$ is the height of the cube, which in this case, are all $3\sqrt{x}$ units), the volume of the cube is equal to $(3\sqrt{x})^3 = 27x\sqrt{x}$ cubic units.

**18. e.** The volume of a rectangular solid is equal to $lwh$, where $l$ is the length of the solid, $w$ is the width of the solid, and $h$ is the height of the solid. If $l$ is the length of solid B and $h$ is the height of solid B, then the length of solid A is $3l$ and the height of solid A is $2h$. Since the volumes of the solids are equal, if $w_1$ represents the width of solid A and $w_2$ represents the width of solid B, then $(3l)(2h)(w_1) = (l)(h)(w_2)$, $6w_1 = w_2$, which means that $w_2$, the width of solid B, is equal to 6 times the width of solid A.

**19. d.** The volume of a cube is equal to the product of its length, width, and height. Since the length, width, and height of a cube are identical in measure, the measure of one edge of Stephanie's cube is equal to the cube root of $64x^6$, which is equal to $4x^2$, since $(4x^2)(4x^2)(4x^2) = 64x^6$. The area of one face of the cube is equal to the product of the length and width of that face. Since every length and width of the cube is $4x^2$, the area of any one face of the cube is $(4x^2)(4x^2) = 16x^4$.

**20. a.** The volume of a rectangular solid is equal to $lwh$, where $l$ is the length of the solid, $w$ is the

width of the solid, and $h$ is the height of the solid. Therefore, $(6)(12)(w) = 36$, $72w = 36$, and $w = \frac{1}{2}$ units.

**21. e.** The surface area of a solid is the sum of the areas of each side of the solid. A rectangular solid has 6 rectangular faces. Two faces measure 4 units by 5 units, two faces measure 4 units by 6 units, and two faces measure 5 units by 6 units. Therefore, the surface area of the solid is equal to $2(4 \times 5) + 2(4 \times 6) + 2(5 \times 6) = 2(20) + 2(24) + 2(30) = 40 + 48 + 60 = 148$ square units.

**22. c.** The volume of a cube is equal to the product of its length, width, and height. Since the length, width, and height of a cube are identical in measure, the measure of one edge of Danielle's cube is equal to the cube root of 512, which is equal to 8, since $(8)(8)(8) = 512$. The area of one face of the cube is equal to the product of the length and width of that face. Since every length and width of the cube is 8 units, the area of one face of the cube is $(8)(8) = 64$ square units. A cube has six faces, so the total surface area of the cube is equal to $(64)(6) = 384$ square units.

**23. b.** The surface area of a solid is the sum of the areas of each side of the solid. A rectangular solid has 6 rectangular faces. If $w$ is the width of the solid, then two faces measure 4 units by 12 units, two faces measure 4 units by $w$ units, and two faces measures 12 units by $w$ units. Therefore, the surface area of the solid is equal to $2(4 \times 12) + 2(4 \times w) + 2(12 \times w) = 96 + 8w + 24w = 96 + 32w$. Since the surface area of the solid is 192 cm$^2$, $96 + 32w = 192$, $32w = 96$, $w = 3$. The width of the solid is 3 units.

**24. e.** The volume of a cube is equal to the product of its length, width, and height. Since the length, width, and height of a cube are identical in measure, the measure of one edge of the cube is equal to the cube root of $x^3$, which is equal to $x$, since $(x)(x)(x) = x^3$. The area of one face of the cube is equal to the product of the length and width of that face. Since every length and width of the cube is $x$, the area of any one face of the cube is $(x)(x) = x^2$. A cube has six faces, so the total surface area of the cube is equal to $6x^2$ square units. It is given that the surface area of the square is $x^3$ square units. Therefore, $6x^2 = x^3$. Divide both sides by $x^2$, and the value of $x$ is 6.

**25. e.** The surface area of a solid is the sum of the areas of each side of the solid. A rectangular solid has 6 rectangular faces. If $x$ is the length of the solid, then $2x$ is the height of the solid and $4x$ is the width of the solid. Two faces of the solid measure $x$ units by $2x$ units, two faces measure $x$ units by $4x$ units, and two faces measures $2x$ units by $4x$ units. Therefore, the surface area of the solid is equal to $2(x \times 2x) + 2(x \times 4x) + 2(2x \times 4x) = 2(2x^2) + 2(4x^2) + 2(8x^2) = 4x^2 + 8x^2 + 16x^2 = 28x^2$.

# ▶ Chapter 15

**1. c.** The circumference of a circle is equal to $2\pi r$, where $r$ is the radius of the circle. Therefore, the circumference of this circle is equal to $(2\pi)(15) = 30\pi$ units.

**2. b.** Angles $AOC$ and $DOB$ are vertical angles; their measures are equal. Therefore, angle $DOB$ is also 60°. The intercepted arc of a central angle is equal in measure to the central angle. Arc $DB$ measures 60°.

**3. a.** The length of arc $CB$ is equal to the size of central angle $COB$ divided by 360, multiplied by the circumference of the circle. Since the radius of the circle is 6 units, the circumference of the circle is $12\pi$; $\frac{100}{360}(12\pi) = \frac{5}{18}(12\pi) = \frac{10}{3}\pi$ units.

**4. d.** The area of a circle is equal to $\pi r^2$; $\pi r^2 = 196\pi$, $r^2 = 196$, $r = 14$. $\overline{CD}$ is a line from one side of the circle, through the center of the circle, to the other side of the circle. It is a diameter, and the measure of a diameter is twice the measure of a radius. The measure of line $CD$ is $(2)(14) = 28$ units.

**5. d.** The area of a circle is equal to $\pi r^2$. The area of this circle is $\pi(8)^2 = 64\pi$ square units. The area of the sector is equal to a fraction of that: $\frac{50}{360}(64\pi) = \frac{5}{36}(64\pi) = \frac{5}{9}(16\pi) = \frac{80}{9}\pi$ square units.

**6. a.** The area of a sector is equal to the area of the circle multiplied by the fraction of the circle that the sector covers. The area of the circle is equal to $\pi(12^2) = 144\pi$. If the area of the sector is equal to $24\pi$, and the angle of the sector is $x$, then $\frac{x}{360}(144\pi) = 24$, $\frac{2x}{5} = 24\pi$, $2x = 120$, $x = 60$. Angle $EOD$ is 60°; therefore, sector $EOD$ is the sector with an area of $24\pi$ square units.

**7. c.** The area of a sector is equal to the area of the circle multiplied by the fraction of the circle that the sector covers. The area of the circle is equal to $\pi(15^2) = 225\pi$. Since the angle of the sector is 40°, $\frac{40}{360}(225\pi) = \frac{1}{9}(225\pi) = 25\pi$ square units.

**8. b.** The length of an arc is equal to the circumference of the circle multiplied by the fraction of the circle that the arc covers. The circumference of the circle is equal to $(2\pi)(27) = 54\pi$ units. The central angle of arc $AE$ is 80°, which means that the length of the arc is $\frac{80}{360}$ of the circumference of the circle: $\frac{80}{360}(54\pi) = \frac{2}{9}(54\pi) = 12\pi$ units.

**9. c.** The length of an arc is equal to the circumference of the circle multiplied by the fraction of the circle that the arc covers. The circumference of the circle is equal to $(2\pi)(9) = 18\pi$ units. The central angle of arc $DB$ is 40°, which means that the length of the arc is $\frac{40}{360}$ of the circumference of the circle: $\frac{40}{360}(18\pi) = \frac{1}{9}(18\pi) = 2\pi$ units.

**10. e.** The area of a circle is equal to $\pi r^2$, where $r$ is the radius of the circle. Therefore, the area of Jasmin's circle is equal to $\pi(9x^2)^2 = (81x^4)\pi$ square units.

**11. e.** The circumference of a circle is $2\pi r$, where $r$ is the radius of the circle. If the circumference of a circle triples, that means the radius of the circle has tripled. The area of the circle has gone from $\pi r^2$ to $\pi(3r)^2 = \pi 9r^2$. The area of the circle is now 9 times bigger.

**12. b.** The area of a circle is equal to $\pi r^2$, where $r$ is the radius of the circle. Therefore, the radius of this circle is equal to the square root of $121x$, or

$11\sqrt{x}$ units. The circumference of a circle is equal to $2\pi r$, so the circumference of this circle is equal to $2\pi(11\sqrt{x}) = (22\sqrt{x})\pi$ units.

**13. d.** The radius of a circle is equal to half the diameter of a circle. The radius of this circle is equal to $\frac{(8x+6)}{2} = 4x + 3$. The area of a circle is equal to $\pi r^2$, where $r$ is the radius of the circle. Therefore, the area of this circle is equal to $\pi(4x+3)^2 = \pi(16x^2 + 12x + 12x + 9) = (16x^2 + 24x + 9)\pi$ square units.

**14. d.** The circumference of a circle is equal to $2\pi r$. Since the radius of a circle is half the diameter of a circle, $2r$ is equal to the diameter, $d$, of a circle. Therefore, the circumference of a circle is equal to $\pi d$. If the diameter is doubled, the circumference becomes $\pi 2d$, or two times its original size.

**15. d.** The area of a circle is equal to $\pi r^2$, where $r$ is the radius of the circle. Therefore, $\pi(2x-7)^2 = (16x+9)\pi$, $4x^2 - 14x - 14x + 49 = 16x + 9$, $4x^2 - 44x + 40 = 0$, $x^2 - 11x + 10 = 0$, $(x-1)(x-10) = 0$, $x = 10$. $x$ cannot be equal to 1, since that would make the radius equal to –5, and a radius cannot have a negative length.

**16. b.** The area of a circle is equal to $\pi r^2$, where $r$ is the radius of the circle. Therefore, the radius of this circle is equal to the square root of $(4x^2 + 20x + 25)$. Factor this trinomial into 2 identical factors; $(2x)(2x) = 4x^2$, $(5)(5) = 25$, and $(2x)(5) + (2x)(5) = 20x$. Therefore, $(4x^2 + 20x + 25) = (2x+5)(2x+5)$, and the square root of $(4x^2 + 20x + 25)$ is $2x + 5$. The diameter of a circle is twice its radius, so the diameter of this circle is $2(2x+5) = 4x + 10$.

**17. e.** The area of a circle is equal to $\pi r^2$, where $r$ is the radius of the circle. Therefore, the area of this circle is $\pi x^2$. The area of a sector is equal to the area of the circle multiplied by the fraction of the circle that the sector covers. That fraction is equal to $\frac{x}{360}$, so the area of the sector is equal to $\frac{x}{360}(\pi x^2) = (\frac{x^3}{360})\pi$ square units.

**18. e.** The length of an arc is equal to the circumference of the circle multiplied by the fraction of the circle that the arc covers. Therefore, if $c$ is the circumference of the circle, then $x\pi = \frac{18}{360}c$. Multiply both sides of the equation by $\frac{360}{18}$, or 20, to isolate $c$: $(\frac{360}{18})x\pi = (\frac{360}{18})\frac{18}{360}c$, $c = 20x\pi$ square units.

**19. e.** Angles $COB$ and $COA$ form a line; they are supplementary angles. Therefore, the measure of angle $COA$ is equal to $180 - 3x$. Since angle $COA$ is a central angle and $CA$ is its intercepted arc, the measure of $CA$ is also $180 - 3x$.

**20. c.** The area of a circle is equal to $\pi r^2$, where $r$ is the radius of the circle. Therefore, the radius of this circle is $\sqrt{12}$ cm or $2\sqrt{3}$ cm. The circumference of a circle is $2\pi r$, so the circumference of this circle is $2\pi(2\sqrt{3}) = 4\sqrt{3}\pi$ cm. A diameter divides a circle into two 180° arcs. Therefore, the measure of arc $AB$ is 180°. The length of the arc is equal to $\frac{180}{360}(4\sqrt{3}\pi) = \frac{1}{2}(4\sqrt{3}\pi) = 2\sqrt{3}\pi$ cm.

**21. d.** The circumference of a circle is $2\pi r$, where $r$ is the radius of the circle, so the radius of this circle is $\frac{16\pi}{2\pi} = 8$ cm. The area of a circle is equal to $\pi r^2$, so the area of this circle is $(8)^2\pi = 64\pi$ cm². The area of a sector is equal to the area of the circle multiplied by the fraction of the circle that the sector covers. That fraction is equal to $\frac{120}{360}$, so the area of the sector is equal to $\frac{120}{360}(64\pi) = \frac{1}{3}(64\pi) = \frac{64}{3}\pi$ cm².

**22. b.** If angles $D$ and $B$ are 70°, then angle $DOB$ is equal to $180 - (70 + 70) = 180 - 140 = 40°$, since there are 180° in a triangle. Angles $DOB$ and $AOC$ are vertical angles; their measures are equal. Since angle $AOC$ is also 40°, and the measure of an intercepted arc of a central angle is equal to the measure of the central angle, arc $AC$ also measures 40°.

**23. c.** The measure of central angle $DOB$ is 60°, since the measure of its intercepted arc, $EF$, is 60°. There are 180° in a triangle, which means that the sum of angles $D$ and $B$ is $180 - 60 = 120$.

Since these angles are equal, each angle measures $\frac{120}{2} = 60°$. Since every angle in the triangle measures 60°, $DOB$ is an equilateral triangle. The radius of the circle is $6x$. $\overline{OF}$ is a radius of the circle, and it is equal in length to $\overline{FB}$. Therefore, the length of side $OB$ of triangle $DOB$ is $2(6x) = 12x$. Since this is an equilateral triangle, all three sides of the triangle measure $12x$. The perimeter of the triangle is $12x + 12x + 12x = 36x$.

**24. c.** The area of a square is equal to the length of one of its sides squared. Therefore, the area of the square is $x^2$ ft.$^2$. The diameter of the circle is equal to the length of a side of the square, $x$, which means that the radius of the circle is equal to $\frac{1}{2}x$ ft. The area of a circle is equal to $\pi r^2$, where $r$ is the radius of the circle. The area of this circle is equal to $(\frac{1}{2}x)^2\pi = (\frac{x^2}{4})\pi$ ft.$^2$. The shaded area is the difference between the area of the square and the area of the circle: $x^2 - (\frac{x^2}{4})\pi = x^2 - \frac{1}{4}x^2\pi$ ft.$^2$.

**25. c.** If the area of the circle is $25\pi$ cm$^2$, then the radius of the circle is 5 cm, since the area of a circle is equal to $\pi r^2$, where $r$ is the radius of the circle. If the radius of the circle is 5 cm, then the diameter of the circle, and the length of a side of the square, is $2(5) = 10$ cm. The diagonal of a square is $\sqrt{2}$ times the length of one of its sides. Therefore, the length of diagonal $AD$ is $10\sqrt{2}$ cm.

**26. b.** If the area of the circle is $8x^2\pi$, then the radius of the circle is $2x\sqrt{2}$, since the area of a circle is equal to $\pi r^2$, where $r$ is the radius of the circle. If the radius of the circle is $2x\sqrt{2}$, then the diameter of the circle, and the length of a side of the square, is $2(2x\sqrt{2}) = 4x\sqrt{2}$. The area of a square is equal to the length of one of its sides squared. Therefore, the area of this square is equal to $(4x\sqrt{2})^2 = 32x^2$. The shaded area is equal to the area of the square minus half the area of the circle: $32x^2 - \frac{(8x^2\pi)}{2} = 32x^2 - 4x^2\pi$.

**27. d.** If the area of the square is 144 square units, then the area of one side of the square is $\sqrt{144} = 12$

units. Since a side of the square is equal to the diameter of the circle, the diameter is 12 units and the radius of the circle is 6 units. The area of the circle is $6^2\pi = 36\pi$ square units. The total area of the figure is equal to the area of the square plus half the area of the circle (since the other half of the circle is within the area of the square): $144 + (\frac{36\pi}{2}) = 144 + 18\pi$ square units.

**28. a.** The length of $\overline{AB}$ is equal to the length of 4 radii of the center circle. Therefore, the radius of the center circle is $\frac{x}{4}$. Since the area of a circle is equal to $\pi r^2$, where $r$ is the radius of the circle, the area of the circle is equal to $\pi(\frac{x}{4})^2 = \frac{(x^2\pi)}{16}$ square units.

**29. a.** If the area of one semicircle is $4.5\pi$ square units, then the area of a whole circle is $2(4.5\pi) = 9\pi$ square units. Since the area of a circle is equal to $\pi r^2$, where $r$ is the radius of the circle, the radius of the circle is equal to $\sqrt{9} = 3$ units. Since the length of $\overline{BC}$ is the length of 2 radii, the length of $\overline{BC}$ is $2(3) = 6$ units. The length of $\overline{AB}$ is the length of 4 radii, or $4(3) = 12$ units. The area of a rectangle is $lw$, where $l$ is the length of the rectangle and $w$ is the width of the rectangle. The area of this rectangle is $(12)(6) = 72$ square units.

**30. b.** The area of the circle is equal to $16\pi$ square units, since the area of a circle is equal to $\pi r^2$, where $r$ is the radius of the circle. The length of $\overline{AB}$ is equal to the length of four radii of the circle placed end to end. Therefore, the length of $\overline{AB}$ is $4(4) = 16$ units. Since $\overline{BC}$ is the length of two radii, its length is $2(4) = 8$ units. The area of a rectangle is $lw$, where $l$ is the length of the rectangle and $w$ is the width of the rectangle. The area of this rectangle is $(16)(8) = 128$ square units. There is one whole circle and two half circles within the rectangle—a total of 2 circles. Subtract the area of the circles from the area of the rectangle to find the size of the shaded area: $128 - 2(16\pi) = 128 - 32\pi$ square units.

# ▶ Chapter 16

**1. b.** The slope of a line is the difference between the y values of two points divided by the difference between the x values of those two points: $\frac{4-6}{7-(-3)}$ $= \frac{-2}{10} = -\frac{1}{5}$.

**2. c.** The slope of a line is the difference between the y values of two points divided by the difference between the x values of those two points: $\frac{-5-(-5)}{-5-5} = \frac{0}{-10} = 0$.

**3. d.** The slope of a line is the difference between the y values of two points divided by the difference between the x values of those two points: $\frac{10-2}{1-(-1)} = \frac{8}{2} = 4$.

**4. a.** The coordinates of point A are (–5,1) and the coordinates of point B are (5,–4). The slope of a line is the difference between the y values of two points divided by the difference between the x values of those two points: $\frac{-4-1}{5-(-5)} = \frac{-5}{10}$ $= -\frac{1}{2}$.

**5. d.** The coordinates of point C are (2,8) and the coordinates of point D are (–1,–7). The slope of a line is the difference between the y values of two points divided by the difference between the x values of those two points: $\frac{-7-8}{-1-2} = \frac{-15}{-3} = 5$.

**6. b.** The midpoint of a line segment is equal to the average of the x values of the endpoints and the average of the y values of the endpoints: $(\frac{0+(-8)}{2}, \frac{-8+0}{2}) = (\frac{-8}{2}, \frac{-8}{2}) = (-4,-4)$.

**7. c.** The midpoint of a line segment is equal to the average of the x values of the endpoints and the average of the y values of the endpoints: $(\frac{6+15}{2}, \frac{-4+8}{2}) = (\frac{21}{2}, \frac{4}{2}) = (10.5,2)$.

**8. a.** The midpoint of a line is equal to the average of the x values of the endpoints and the average of the y values of the endpoints: $(\frac{0+0}{2}, \frac{-4+4}{2}) = (\frac{0}{2}, \frac{0}{2}) = (0,0)$.

**9. d.** The coordinates of point A are (–5,–4) and the coordinates of point B are (9,2). The midpoint of a line is equal to the average of the x values of the endpoints and the average of the y values of the endpoints: $(\frac{-5+9}{2}, \frac{-4+2}{2}) = (\frac{4}{2}, \frac{-2}{2}) = (2,-1)$.

**10. a.** The coordinates of point C are (–3,6) and the coordinates of point D are (5,–6). The midpoint of a line is equal to the average of the x values of the endpoints and the average of the y values of the endpoints: $(\frac{-3+5}{2}, \frac{6+(-6)}{2}) = (\frac{2}{2}, \frac{0}{2}) = (1,0)$.

**11. c.** To find the distance between two points, use the distance formula:
$$D = \sqrt{((x_2 - x_1)^2 + (y_2 - y_1)^2)}$$
$$D = \sqrt{((2-(-6))^2 + (17-2)^2)}$$
$$D = \sqrt{(8^2 + (15)^2)}$$
$$D = \sqrt{(64 + 225)}$$
$$D = \sqrt{289} = 17 \text{ units}$$

**12. e.** To find the distance between two points, use the distance formula:
$$D = \sqrt{((x_2 - x_1)^2 + (y_2 - y_1)^2)}$$
$$D = \sqrt{((4-0)^2 + (4-(-4))^2)}$$
$$D = \sqrt{(4^2 + 8^2)}$$
$$D = \sqrt{(16 + 64)}$$
$$D = \sqrt{80} = \sqrt{16}\sqrt{5} = 4\sqrt{5} \text{ units}$$

**13. d.** To find the distance between two points, use the distance formula:
$$D = \sqrt{((x_2 - x_1)^2 + (y_2 - y_1)^2)}$$
$$D = \sqrt{((7-3)^2 + ((-6)-8)^2)}$$
$$D = \sqrt{(4^2 + (-14)^2)}$$
$$D = \sqrt{(16 + 196)}$$
$$D = \sqrt{212} = \sqrt{4}\sqrt{53} = 2\sqrt{53} \text{ units}$$

**14. d.** The coordinates of point A are (–5,–4) and coordinates of point B are (9,2). To find the distance between two points, use the distance formula:
$$D = \sqrt{((x_2 - x_1)^2 + (y_2 - y_1)^2)}$$
$$D = \sqrt{((9-(-5))^2 + (2-(-4))^2)}$$
$$D = \sqrt{(14^2 + (6)^2)}$$
$$D = \sqrt{(196 + 36)}$$
$$D = \sqrt{232} = \sqrt{4}\sqrt{58} = 2\sqrt{58} \text{ units}$$

**15. d.** The coordinates of point C are (–3,6) and the coordinates of point D are (5,–6). To find the distance between two points, use the distance formula:
$$D = \sqrt{((x_2 - x_1)^2 + (y_2 - y_1)^2)}$$
$$D = \sqrt{((5-(-3))^2 + ((-6)-6)^2)}$$
$$D = \sqrt{(8^2 + (-12)^2)}$$

$$D = \sqrt{(64 + 144)}$$
$$D = \sqrt{208} = \sqrt{16}\sqrt{13} = 4\sqrt{13} \text{ units}$$

**16. c.** Divide both sides of the equation by 5 to put the equation in slope-intercept form ($y = mx + b$); $5y = -3x + 6$ is equivalent to $y = -\frac{3}{5}x + \frac{6}{5}$. The slope of this line is the coefficient of $x$, $-\frac{3}{5}$.

**17. a.** Parallel lines have identical slopes. When an equation is written in slope-intercept form ($y = mx + b$), the slope of the line is $m$. The slope of the line $y = -2x + 4$ is –2. Only choice **a** is a line with a slope of –2.

**18. e.** The slopes of perpendicular lines are negative reciprocals of each other. When an equation is written in slope-intercept form ($y = mx + b$), the slope of the line is $m$. The slope of the line $y = -\frac{1}{6}x + 8$ is $-\frac{1}{6}$. The negative reciprocal of $-\frac{1}{6}$ is 6. Only choice **e** is a line with a slope of 6.

**19. d.** Divide both sides of the equation by 4 to put the given equation in slope-intercept form ($y = mx + b$); $4y = 6x - 6$ is equivalent to $y = \frac{3}{2}x - \frac{3}{2}$. The slope of this line is the coefficient of $x$, $\frac{3}{2}$. Parallel lines have identical slopes. Only choice **d** is a line with a slope of $\frac{3}{2}$.

**20. b.** Divide both sides of the given equation by –2 to put the equation in slope-intercept form ($y = mx + b$); $-2y = -8x + 10$ is equivalent to $y = 4x - 5$. The slope of this line is the coefficient of $x$, 4. The slopes of perpendicular lines are negative reciprocals of each other. The negative reciprocal of 4 is $-\frac{1}{4}$. Only choice **b** is a line with a slope of $-\frac{1}{4}$.

**21. e.** To find the distance between two points, use the distance formula:
$$D = \sqrt{((x_2 - x_1)^2 + (y_2 - y_1)^2)}$$
$$D = \sqrt{((x - (-x))^2 + ((-y) - y)^2)}$$
$$D = \sqrt{((2x)^2 + (-2y)^2)}$$
$$D = \sqrt{(4x^2 + 4y^2)}$$
$$D = \sqrt{4(x^2 + y^2)}$$
$$D = 2\sqrt{(x^2 + y^2)}$$

**22. c.** The slopes of perpendicular lines are negative reciprocals of each other. The slope of the other line is $-\frac{1}{3}$. The equation of the other line is $y = -\frac{1}{3}x + b$, where $b$ is the $y$-intercept of the line.

Use the point (1,5) to find the $y$-intercept; $5 = -\frac{1}{3}(1) + b$, $b = \frac{16}{3}$. Therefore, the equation of the line is $y = -\frac{1}{3} + \frac{16}{3}$.

**23. c.** The midpoint of a line is equal to the average of the $x$ values of the endpoints and the average of the $y$ values of the endpoints: $(\frac{2x + 3 + 10x - 1}{2}, \frac{y - 4 + 3y + 6}{2}) = (\frac{12x + 2}{2}, \frac{4y + 2}{2}) = (6x + 1, 2y + 1)$.

**24. a.** The slopes of perpendicular lines are negative reciprocals of each other. Therefore, if the slope of one line is $m$, the slope of the other line is $-\frac{1}{m}$. The product of these slopes is $(m)(-\frac{1}{m}) = -1$.

**25. d.** The slopes of perpendicular lines are negative reciprocals of each other. Therefore, if the slope of line $A$ is $m$, the slope of line $B$ is $-\frac{1}{m}$. If the slope of line $A$ is multiplied by 4, it becomes $4m$. The negative reciprocal of $4m$ is $-\frac{1}{4m}$. To change the original slope of line $B$, $-\frac{1}{m}$, to this new slope, $-\frac{1}{4m}$, you must multiply by $\frac{1}{4}$.

## ▶ Posttest

**1. e.** Factor the numerator and denominator; $(x^2 + 4x - 12) = (x - 2)(x + 6)$ and $(x^2 - 8x + 12) = (x - 2)(x - 6)$. The term $(x - 2)$ is common to the numerator and denominator, so it can be canceled. This leaves $\frac{x + 6}{x - 6}$.

**2. c.** The length of an arc is equal to the circumference of the circle, $2\pi r$, multiplied by the measure of the angle formed by the two radii that intercept the arc, divided by 360: $2\pi(25)(\frac{72}{360}) = 50\pi(\frac{1}{5}) = 10\pi$.

**3. b.** Graph the line $y = 1$. The graph of this line crosses the graphed equation in 8 places. Therefore, there are at least 8 different values for which the function, $f(x)$, is equal to 1. The function could contain more than 8 values for which $f(x) = 1$ if more of the coordinate plane was visible. Since the graphed equation passes the vertical line test, it is a function. The function extends beyond the line $x = 4$; therefore, there are values greater than 4 in the domain of the function.

ANSWERS

There are many $y$ values between $-2$ and 1, as the range of the function shown extends from $-3$ to 4. The graphed equation crosses the $y$-axis at only one point, $(0,3)$, so the equation has only one $y$-intercept.

**4. b.** To find the midpoint of the line segment, find the average of the $x$-coordinates and the average of the $y$-coordinates of the endpoints of the line segment: $\frac{(-1+13)}{2} = \frac{12}{2} = 6; \frac{(4+12)}{2} = \frac{16}{2} = 8$. The midpoint of the line is $(6,8)$.

**5. d.** If the numerator of a fraction is doubled, the denominator of the fraction must be doubled in order for the value of the fraction to remain the same; $\frac{x}{yz} = \frac{2x}{2yz}$. The denominator, $2yz$, represents $(2y)z$ or $(2z)y$. Either the value of $y$ is doubled, or the value of $z$ is doubled. Notice that both values are not doubled. If that were the case, the denominator would be equal to $(2y)(2z) = 4yz$, and the value of the fraction would be halved.

**6. e.** When multiplying like bases, keep the base and add the exponents: Since $1 + (-2) + 3 = 2, a^1a^{-2}a^3 = a^2$. Since $-1 + 2 + (-3) = -2, b^{-1}b^2b^{-3} = b^{-2}$; $(a^2)(b^{-2}) = \frac{(a^2)}{(b^2)}$.

**7. a.** Cross multiply: $(a+5)(a-2) = (3)(4a), a^2 + 3a - 10 = 12a, a^2 - 9a - 10 = 0$. Factor this quadratic and solve for $a$: $(a+1)(a-10) = 0, a+1 = 0, a = -1; a - 10 = 0, a = 10$.

**8. d.** Plug each answer choice into the inequality; $4(6) + 6 > 3(4) + 15, 24 + 6 > 12 + 15, 30 > 27$. Since the inequality $30 > 27$ is true, choice **d** is correct.

**9. d.** The area of a circle is equal to $\pi r^2$, where $r$ is the radius of the circle. Since the radius of this circle is $r$, the area of this circle is $\pi r^2$. The area of a square is equal to the length of one side squared. The length of a side of the square is equal to the diameter of the circle, which is $2r$. Therefore, the area of the square is equal to $(2r)(2r) = 4r^2$. The difference between the area of the square and the area of the circle is $4r^2 - \pi r^2$. This area represents the area between the circle and the square. Since only one of these

four regions is shaded, divide the difference of the areas by 4: $\frac{(4r^2 - \pi r^2)}{4} = r^2 - \frac{\pi r^2}{4}$.

**10. d.** Angles $a$, $b$, and $c$ form a line and angles $d$, $e$, and $f$ are the angles of a triangle. Since there are 180° in a line and 180° in a triangle, the sum of $a$, $b$, and $c$ and the sum of $d$, $e$, and $f$ are both 180°. Since $a + b + c = 180, a + c = 180 - b$. Angles $b$ and $e$ are vertical angles, so their measures are equal. Therefore, $a + c = 180 - e$. Also, since $a + b + c = 180$ and $b = e, a + e + c = 180$. And, since $d + e + f = 180, a + b + c + d + e + f = 360$. However, $b + c$ is not equal to $e + f$. Although $b = e, c$ is not equal to $f$; $a = f$ and $c = d$, since those are pairs of alternating angles.

**11. e.** First, rewrite the equation in slope-intercept form $(y = mx + b); 4y + 3x = 12, 4y = -3x + 12, y = -\frac{3}{4}x + 3$. The slopes of perpendicular lines are negative reciprocals of each other. Therefore, the slope of a line perpendicular to $y = -\frac{3}{4}x + 3$ is $\frac{4}{3}x$, since $\frac{4}{3}$ is the negative reciprocal of $-\frac{3}{4}$. The only line given that has a slope of $\frac{4}{3}$ is choice **e**.

**12. c.** First, express the area of the original circle in terms of $g$. The circumference of a circle is equal to $2\pi r$, where $r$ is the radius of the circle. Therefore, $g = 2\pi r$. The radius in terms of $g$ is equal to $\frac{g}{2\pi}$. The area of a circle is equal to $\pi r^2$. Replace $r$ with $\frac{g}{2\pi}$. The area of the circle is equal to: $\pi(\frac{g}{2\pi})^2 = \frac{g^2}{4\pi}$. Since the area is tripled, the new area of the circle is equal to 3 times this quantity: $\frac{3g^2}{4\pi}$.

**13. a.** To find the turning point of a parabola, find the value that makes the $x$ term of the equation equal to 0. Then, use that value of $x$ to find the value of $y$. The $x$ term of $y = (x-2)^2 - 2$ will be 0 when $x = 2$, since $(2-2)^2 = 0$. The $y$-coordinate of that point is equal to $(2-2)^2 - 2 = -2$, making the coordinates of the turning point $(2,-2)$. Since the turning point of $y = x^2$ is $(0,0)$, the turning point of the graph of $y = (x-2)^2 - 2$ (and therefore, the entire graph) has been shifted 2 units right and 2 units down, relative to the graph of $y = x^2$.

**14. c.** If the tangent of an angle in a right triangle is 1, that means the lengths of the bases of the triangle are equal, since the tangent is equal to the length of the opposite base divided by the length of the adjacent base. Therefore, triangle *ABC* is an isosceles right triangle. The hypotenuse of an isosceles right triangle is equal to $\sqrt{2}$ times the length of one of the bases. Since the length of each base is 10 units, the length of $\overline{AC}$ is $10\sqrt{2}$ units.

**15. b.** Divide each term in the numerator by $3x^2$. Divide the coefficient of each term by 3, and subtract 2 from the exponent of each $x$ term. $\frac{12x^4}{3x^2} = 4x^2$, $\frac{21x^3}{3x^2} = 7x$, $\frac{-3x^2}{3x^2} = -1$. Therefore, $\frac{12x^4 + 21x^3 - 3x^2}{2} = 4x^2 + 7x - 1$.

**16. 6** The area of a sector of a circle is equal to the area of a circle, $\pi r^2$, multiplied by the angle formed by the two radii of the sector divided by 360. Therefore, $8\pi = (\frac{80}{360})\pi r^2$, $(\frac{36}{8\pi})(8\pi) = r^2$, $36 = r^2$, $r = 6$.

**17. 8** The area of a triangle is equal to $\frac{1}{2}bh$, where $b$ is the base of the triangle and $h$ is the height. If each side of *ABC* is three times the length of its corresponding side of *DEF*, then the area of *ABC*, relative to *DEF*, is $\frac{1}{2}(3b)(3h) = \frac{9}{2}bh$, or nine times the size of the area of *DEF* ($\frac{1}{2}bh$). Therefore, the area of *DEF* is equal to one-ninth of the area of *ABC*: $\frac{72}{9} = 8$ square units.

**18. 16** First, take the cube (third) root of both sides of the given equation. The cube root of $a^{\frac{3}{2}} = a^{\frac{1}{2}}$ and the cube root of 512 is equal to 8. Since $a^{\frac{1}{2}} = 8$, square both sides of the equation to find the value of $a$: $a^{(\frac{1}{2})(2)} = 8^2$, $a = 64$. Now, substitute the value of $a$ into the second expression; $a^{\frac{2}{3}} = 64^{\frac{2}{3}}$. The cube root of 64 is 4, since $(4)(4)(4) = 64$. The square of 4 is 16; therefore, $64^{\frac{2}{3}} = 16$.

**19. 210** The area of a rectangle is equal to its length multiplied by its width. Therefore, if the length of the court is 80 and the area of the court is 2,000, then the width of the court is equal to: $\frac{2,000}{80} = 25$ ft. The perimeter of a rectangle is equal to twice its length plus twice its width. Therefore, the perimeter of the court is equal to: $(2)(80) + (2)(25) = 160 + 50 = 210$ ft.

**20. 32** First, find the volume of one book. The volume of a rectangular prism is equal to *lwh*, where *l* is the length of the prism, *w* is the width, and *h* is the height. The volume of one book is equal to: $(8)(5)(1) = 40$ cubic inches. The volume of the box is equal to $(16)(10)(8) = 1,280$ cubic inches. Divide the volume of the box by the volume of one book to find how many books can fit in the box: $\frac{1,280}{40} = 32$.

**21. 10.5** Substitute $-4$ for $x$: $\frac{(-4)^2 + 3(-4)}{8 - 4} = \frac{16 - 12}{8 - 4} = 12 - 1.5 = 10.5$.

**22. 14** If a radius is drawn to the point where a tangent touches a circle, a right angle is formed. Since angle *AOB* is a central angle, its measure is equal to the measure of its intercepted arc. Therefore, the measure of angle *AOB* is 60°. Radius *AO*, secant *EO*, and segment *EA* of tangent *CD* form a right triangle. The cosine of angle *O* is equal to the adjacent side, *AO*, divided by the hypotenuse, *EO*. Since the cosine of $60° = \frac{1}{2}$, $\frac{1}{2} = \frac{7}{\overline{EO}}$, $(\frac{1}{2})\overline{EO} = 7$, $\overline{EO} = 14$. Alternatively, you can recognize right triangle *EAO* is a 30-60-90 right triangle. Since $\overline{AO}$ is opposite the 30° angle of the triangle, it is the shorter base of the triangle. The hypotenuse of a 30-60-90 right triangle is twice the length of the shorter base, so the length of *EO* is equal to $(2)(7) = 14$.

**23. 2** Use the first equation to write $a$ in terms of $b$: $-3a - 9b = -6$, $-3a = 9b - 6$, $a = -3b + 2$. Substitute this value for $a$ in the second equation: $5(-3b + 2) + 6b = -8$, $-15b + 10 + 6b = -8$, $-9b = -18$, $b = 2$.

**24.** **384** Each term is twice the previous term. Therefore, the fifth term in the pattern is $(48)(2) = 96$, the sixth term is $(96)(2) = 192$, and the seventh term is $(192)(2) = 384$. Alternatively, each term is equal to 6 times a power of 2. The first term is equal to $(6)(2^0) = 6$, the second term is equal to $6(2^1) = 12$. The exponent of 2 is equal to one less than the position of the term in the sequence. Therefore, the seventh term is equal to $(6)(2^{7-1}) = (6)(2^6) = (6)(64) = 384$.

**25.** **3** The expression is undefined when the denominator is equal to 0. Factor the denominator to find the values of $x$ that make the denominator equal to 0; $(x^2 + 6x - 27) = (x + 9)(x - 3)$; $x + 9 = 0$, $x = -9$; $x - 3 = 0$, $x = 3$. The positive value of $x$ that makes the expression undefined is 3.

**NOTES**

# NOTES

**NOTES**

# NOTES

# Special FREE Online Practice from LearningExpress!

**Let LearningExpress help you acquire practical, essential algebra and geometry skills FAST**

Go to the LearningExpress Practice Center at www.LearningExpressFreeOffer.com, an interactive online resource exclusively for LearningExpress customers.

Now that you've purchased LearningExpress's *411 SAT Algebra and Geometry Questions*, you have **FREE** access to:

- **54 practice questions** covering **ALL** types of **math questions** found on the **SAT** including **algebra I and II, geometry, arithmetic, and more**
- **Immediate scoring** and **detailed answer explanations**
- Benchmark your skills and focus your study with our **customized diagnostic report**
- **Improve** your SAT math knowledge and **learn** how to work each type of SAT math problem

Follow the simple instructions on the scratch card in your copy of *411 SAT Algebra and Geometry Questions*. Use your individualized access code found on the scratch card and go to www.LearningExpressFreeOffer.com to sign in. Start practicing your SAT math skills online right away!

Once you've logged on, use the spaces below to write in your access code and newly created password for easy reference:

Access Code: _____     Password: _____